THE SHAPE OF SPEED

STREAMLINED AUTOMOBILES AND MOTORCYCLES, 1930–1942

THE SHAPE of SPEED

STREAMLINED AUTOMOBILES AND MOTORCYCLES, 1930–1942

Ken Gross
David Rand
Peter Harholdt, principal photographer

Portland Art Museum

Published in conjunction with the exhibition "The Shape of Speed: Streamlined Automobiles and Motorcycles 1930–1942,"
which was organized by the Portland Art Museum, Portland, Oregon, with Guest Curator Ken Gross, June 16-September 16, 2018

Text Copyright © 2018 by Ken Gross, unless otherwise noted
Compilation Copyright © 2018 by the Portland Art Museum

Published by:

St. Paul, Minnesota
www.stanceandspeed.com

All rights reserved. With the exception of quoting brief passages for the purposes of review, no part
of this publication may be reproduced without prior written permission from the publisher.

The information in this book is true and complete to the best of our knowledge. All recommendations are made
without any guarantee on the part of the author or publisher, who also disclaim any liability incurred in connection
with the use of this data or specific details.

We recognize, further, that some words and designations mentioned herein are the property of the trademark holder.
We use them for identification purposes only. This is not an official publication.

Published June 2018
Printed in China

ISBN: 978-0-9891149-8-1
Library of Congress Control Number: 2018936050
Gross, Ken

The Shape of Speed: Streamlined Automobiles and Motorcycles 1930–1942
Edited by Peter Bodensteiner
Layout and Design by John Sticha
Proofreading by Ian Gillingham
Printer Liaison Services by Jim Bindas

Front Cover: 1934 Graham Combination Coupe, Peter Harholdt
Frontispiece: 1939 Steyr 55 "Baby" Coupe, Peter Harholdt
Title Page: 1942 Alfa Romeo 6C 2500 SS Bertone Berlina, Carrstudio/Collezione Lopresto
Contents Page and Section Break Part I: 1938 Tatra T77A, Peter Harholdt
Director's Foreword: 1938 Mercedes-Benz Mercedes-Benz 540K Stromlinienwagen in wind tunnel, Daimler AG
Section Break, Part II: 1934 Chrysler Imperial Model CV Airflow Coupe, Peter Harholdt
Final Spread: 1936 Cord 812SC Westchester Sedan, Peter Harholdt
Back Page: 1936 Stout Scarab, Peter Harholdt

CONTENTS

DIRECTOR'S FOREWORD AND ACKNOWLEDGMENTS

In these pages, we celebrate the design that moves us. "The Shape of Speed: Streamlined Automobiles and Motorcycles, 1930–1942" continues the Portland Art Museum's exhibition program that presents uniquely designed objects as works of art. We first introduced our visitors to automotive design with the 2011 exhibition, "The Allure of the Automobile."

"The Shape of Speed" takes a closer look at just over a decade of streamlining—a concept that has fascinated people for generations. Beginning in the 1930s, just as the Great Depression took hold, designers embraced the challenge of styling and building truly streamlined vehicles to inspire a sense of hope for the future. Through the European and American automobile and motorcycle examples we have assembled in Portland, the exhibition demonstrates how designers translated the concept of streamlining into exciting machines that look as though they were moving while at rest.

We are indebted to the lenders of the automobiles and motorcycles listed on page 108 for their incredible generosity in sharing their exceptional objects with our community.

This exhibition would not have been possible without the advocacy, support, and counsel of our Shape of Speed Society Honorary Chairs, which include Robert and Kathleen Ames, Kevin Blount, Keith Martin, and Jim Mark. We are forever grateful for their passionate embrace of this exhibition and the Museum.

Underwriting the unique expenses of bringing seventeen automobiles and two motorcycles to Portland and into our galleries from throughout the United States and Europe are our generous donors. They include: The Standard; Daimler Trucks North America; The Melvin Mark Companies; Nani S. Warren and The Swigert Warren Foundation; Helen Jo and William Whitsell; Alfa Romeo and Fiat Chrysler Automobiles; Sports Car Market 30th Anniversary Tour; Kathleen and Robert Ames; Mr. Ken Austin; Jerry Baker and Janet Geary; Kevin Blount; Bonhams; Mary Beth and Roger Burpee; Lana and Christian Finley; Eric and Jan Hoffman; Mr. Keith Martin; Laura Meier; Merrill Lynch Private Banking & Investment Group; Mr. Mark J. and Dr. Jennifer Miller; Mark and Katherine Frandsen; Elizabeth Lilley; and Jim and Susan Winkler. Additional support was provided by the James F. and Marion Miller Foundation, Meyer Memorial Trust, William G. Gilmore Foundation, and the Exhibition Series Sponsors.

We are especially grateful for the passion and knowledge of the exhibition's guest curator, Ken Gross. At the Portland Art Museum, I would like to acknowledge director of collections and exhibitions Donald Urquhart for overseeing the logistics of the project; chief preparator Matthew Juniper for presenting the objects as works of art; and associate registrar Amanda Kohn for coordinating their safe transport. I would also like to thank director of development Karie Burch and assistant to the director Elizabeth Thomas for providing key support throughout the project as well as curatorial associate for collections and exhibitions Charles Campbell and catalogue publisher Peter Bodensteiner for creating this beautiful publication to commemorate the exhibition. With this new exhibition, the Museum opens the door to another significant era and exploration of design.

Brian J. Ferriso
The Marilyn H. and Dr. Robert B. Pamplin Jr. Director
Chief Curator
Portland Art Museum

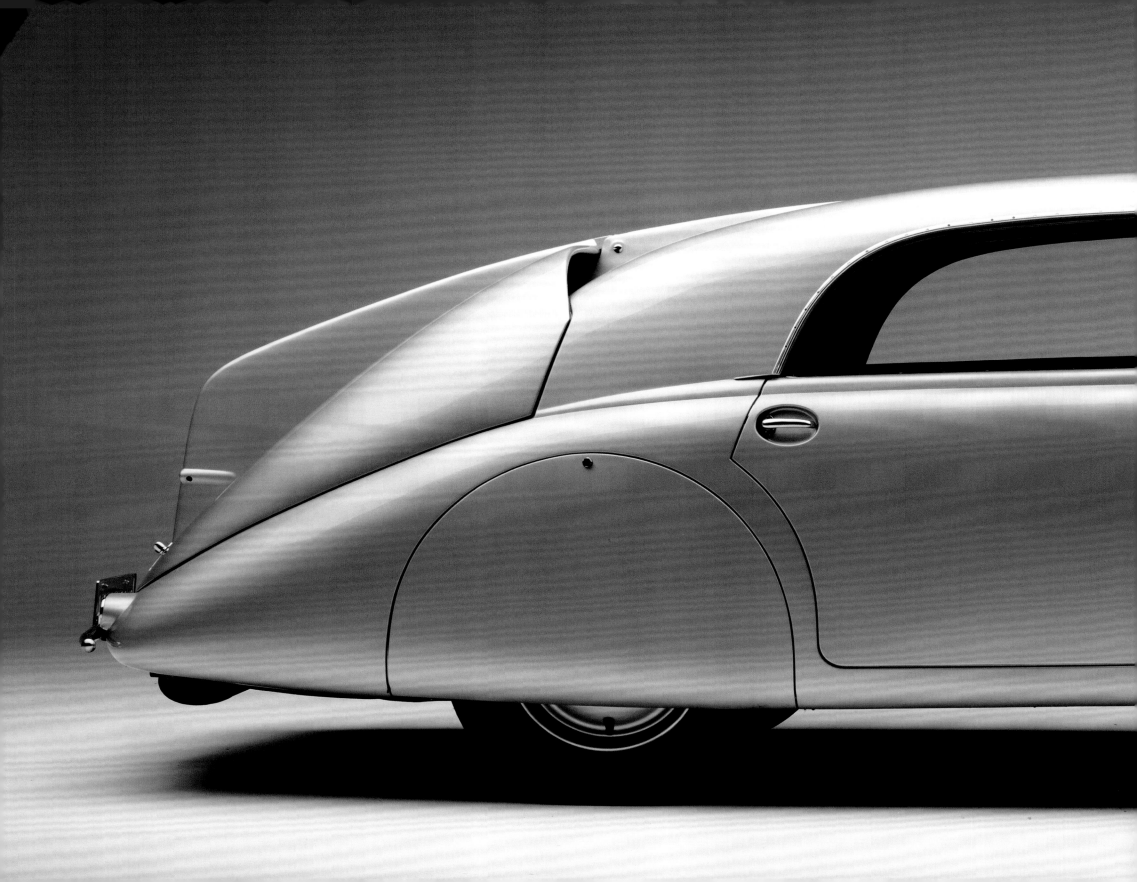

THE STREAMLINE VOGUE

THE INFLUENCE OF SCIENCE, ART AND FASHION ON AUTOMOTIVE DESIGN: 1930 TO 1942

by David Rand

By the 1930s, automotive design had reached a point where new manufacturing techniques, advancing material technologies, and a growing public desire for something more than just transportation were beginning to have major impact on appearance. The resulting period was marked by some of the most outstanding and memorable designs of the classic era—vehicles that today continue to be highly prized by collectors. Among the most compelling of these were the streamline designs of the time. Expressing a visual dynamism and technical modernism, these vehicles, despite being produced during the Depression years (or perhaps *because* of the era's challenges), suggested a more optimistic future where aesthetics and science could be perfectly compatible. Yet, while these cars embraced the *appearance* of aerodynamics, in most cases there was little reality behind this effort, despite there having been attempts to optimize vehicle aerodynamics going back to the beginning of the century.

Early Aerodynamic Development

Designers of early competition cars were then, as now, looking for ways to achieve greater speed and efficiency. It was understood at the time that the reduction of air resistance would be one way to achieve this advantage. Despite the first wind tunnel having been created in 1871 (by Francis Wenham of Great Britain), there was little aerodynamic information available. Most of the designs of these vehicles relied more on intuition than science. In 1902, Walter Baker, the founder of the Baker Motor Vehicle Company, was the first man to break the 100-mph mark in a car. The Baker Torpedo, a design inspired by the ovoid form that Baker observed in a falling drop of oil, was remarkable for its time.

Its appearance was closer to that of a 1950s Bonneville belly tank racer than anything automotive in its period. In 1906, a Stanley Steamer set the world record for the fastest mile in an automobile with a low, tapered body that resembled an inverted canoe, and that clearly took air resistance into consideration. The 1910 Buick Bug used a unique cooling system that allowed it to reduce the large frontal area normally necessitated by large radiators. And by 1917, Harry Miller's *Golden Submarine* (Fig. 1), which employed an enclosed cabin and a distinctively teardrop form, was claimed to have been developed with the help of a wind tunnel, and as such would be one of the first streamlined race cars of its kind.

Fig. 1: Barney Oldfield with the *Golden Submarine* streamlined race car. Photo by ISC Images & Archives via Getty Images

Non-competitive vehicles also experimented with early forms of aerodynamics. Edmund Rumpler was a manufacturer of aircraft until 1921 when he introduced the Tropfenwagen ("drop car," as in water drop) (Fig. 2). The vehicle featured a smooth undertray and mid-mounted engine, horizontal mudguards, the first use of curved glass, and a distinctively teardrop plan-view form. Rumpler's vehicle was remarkably aerodynamically efficient for its time, and the shape's effectiveness was confirmed by wind tunnel testing in 1979. Fewer than 100 examples were built. Starting in 1927, Sir Charles Burney, who also had an aircraft-design background, built a series of vehicles that similarly utilized a teardrop form. In this case Burney mounted the engine in the long rear overhang that the teardrop proportions imposed. But with only a few made, and an unorthodox appearance, both the Rumpler and the Burney cars would remain curiosities more than indicators of the automotive future.

Paul Jaray (1889-1974) was perhaps the man most responsible for establishing real and practical aerodynamic applications for the automobile. An Austrian who studied aeronautical engineering, he was, by the beginning of World War I, working for the Luftschiffbau Zeppelin Company. There he was instrumental in developing airships based on the simple cigar shape used before the war, followed by the tapered teardrops that would provide the basis for the Graf Zeppelin and later, the Hindenburg. By the end of the war, while still at Zeppelin, Jaray turned his attention to automotive applications. By 1921 he had filed for patents that applied aerodynamic principles to road vehicles. His early patent applications illustrated designs not unlike his aircraft, and they were remarkably advanced for the time. They incorporated continuously tapering teardrop forms, integrated fenders, uppers blended into the lower body with recessed wheels, and a single centered headlamp and radiators.

By 1923 Jaray would build a series of prototypes to promote his ideas (Fig. 3). By 1927, having been awarded his patents, he would form the Stromlinien Karosserie Gesellschaft (Streamline Coachwork Company) in Zurich, with a sister branch in the US in 1931 called the Streamline Corporation. (Curiously, the terms "streamline" and "aerodynamic"

Fig. 2: 1921 Rumpler Tropfenwagen "Drop Car." Photo by Bob D'Olivo/The Enthusiast Network/Getty Images

would be used interchangeably at this time). Despite Jaray's advanced ideas, they met with little commercial success in the 1920s. Jaray's early prototypes were based on typical chassis of the day and thus were awkwardly tall and narrow, and aesthetically challenging. While other manufacturers would eventually build prototypes based on Jaray's patents, it would not be until the next decade when the promise of his ideas would find real-world, if somewhat limited, application.

The Impact of the Industrial Designer

By the early 30s, a relatively new profession was having a major influence on product development, especially in the United States. Manufacturers were now more willing to rely on the industrial designer, who combined the skills of aesthetic judgment with the pragmatic needs of manufacturing, to add freshness and visual excitement

to consumer products. Given the depressed economy of the period, in order to maintain sales, producers of home goods and appliances wanted designs that would leverage newly available materials and manufacturing techniques. Designers such as Raymond Loewy, Walter Dorwin Teague, Norman Bel Geddes, and Henry Dreyfuss would become established during this period. And it was these men, with their rejection of the previous "decorative" design idiom in favor of the new "machine age" aesthetic, who would promote this new approach through the embrace of streamlining. Products such as the classic Loewy pencil sharpener (Fig. 4) and the Dreyfuss-designed Hoover vacuum cleaner exhibit a dominant teardrop form, which was perceived as the theoretically perfect aerodynamic shape. Unlike the early automotive aerodynamic studies that utilized the same form for its functional attributes, however, the shapes of these

products had little to do with their function. Indeed, one of the hallmarks of streamlining was its ability to project a sense of the "modern," and the "new," upon almost any product. Still, by its very nature, the streamline aesthetic would be best served when applied to dynamic objects.

In this period, the railroads had discovered that by "shrouding" old steam engines with sleek new bodies, they achieved incremental performance gains. More importantly, as designer Otto Kuhler noted in a 1935 speech, "the man on the street" couldn't distinguish a 1915 locomotive from a 1935 locomotive. To the public these old, facelifted engines became distinctive and new again. This allowed the railroads to avoid the risk inherent in adopting the still-developing diesel engine technology, as well as the investment expense of building an all-new engine. In 1934, the Union Pacific railroad would introduce the *M-10000* passenger train

Fig. 3: Experimental streamlined cars designed by aerodynamicist Paul Jaray in Berlin, circa 1935. Photo by Keystone/Getty Images

Fig. 4: Raymond Loewy-designed prototype pencil sharpener, 1933. Private Collection/Photo © Christie's Images/Bridgeman Images

Fig. 5: 1934 Chrysler Imperial Airflow and Pullman *M-10000*. Courtesy of FCA North America

(formerly the Aquatic Park Casino) of San Francisco (Fig. 8 and 9), went one step further by taking on a distinctively nautical theme, complete with portholes and an upper bridge structure in the case of the Coca-Cola plant.

The Convergence of Aerodynamics and Streamlining

During the 1930s, the streamline trend had become ubiquitous in many forms of product design, and the automobile was no exception. Indeed few products lent themselves better to the emerging fashion. At the 1933-34 Century of Progress in Chicago, automobile manufacturers displayed "dream" cars that demonstrated the varying and evolving state of streamline-inspired design. General Motors displayed the Cadillac Aero-Dynamic coupe, which featured a smooth fastback profile with a recessed rear license plate, and a spare tire hidden within the trunk, which was unusual at this time. Pontoon fenders, a fast windshield angle, and distinctive, horizontal, airfoil-shaped hood louvers would reinforce the contemporary, if measured, streamlined look. Cadillac would produce a production version of the vehicle shortly afterward. Despite naming the vehicle "Aero-Dynamic," it is doubtful that any aerodynamic testing or evaluation was done on this one-off show car. In a letter to stockholders at the time, GM Chairman Alfred P. Sloan stated that the company's 1935 models would show "the trend toward streamlining—vogue very much in evidence at this time ... ," though, "the contribution of streamlining is definitely limited to the question of styling," and that any belief that it led to greater efficiencies was a "popular misunderstanding."[2]

In comparison, the Pierce-Arrow Silver Arrow was more radical. Phillip Wright's very advanced design included flush front fenders that were integrated with the body, and which also cleverly hid the twin spare tires. The Silver Arrow also featured integrated headlamps—a Pierce-Arrow hallmark since 1913—a laid-back windshield and grille, skirted rear wheels, and a tapered rear profile. In short it incorporated all the visual elements that would come to signify automotive streamline styling (Fig. 7). An even more daring design was shown at Ford, called the Briggs dream car. Because Ford would not have its own in-house design department until

(Fig. 5), while the Burlington railroad would launch the *Burlington Zephyr*, both using the new diesel engines and both clearly using the streamline aesthetic. Dreyfuss and Loewy would also design shrouds for locomotives, with each designer showing a distinctive style. The railroads would also eventually build new steam-powered locomotives using this aerodynamic approach, and as a group, all of these engines would be referred to as "streamliners."[1]

Designers would also experiment with streamlining in nautical applications. In 1933, Loewy redesigned an older ferry for the Virginia Ferry Corporation, rechristened *Princess Anne*. It featured a superstructure that was distinctively rounded, with paint graphics emphasizing its linear quality. The *Kalakala*, another rebuilt ferry operated

out of Seattle beginning in 1935, may have been a novelty when introduced (Fig. 6). There was never a fleet of these boats, and it's questionable how much the shape actually helped performance at the speed it was operated. But it was a unique-looking vessel, clearly designed to make a statement, which, like the streamlined locomotives, was really the primary purpose.

Architecture would also feel the impact of streamlining, in the form of *Streamline Moderne*. Closely related to the Art Deco movement and often mistaken for it, *Streamline Moderne* emphasized the horizontal and incorporated rounded corners that suggested motion, as in other streamlined objects. Some, like the Coca-Cola bottling plant located in Los Angeles, and the Maritime Museum

1935, the company relied on its body supplier, Briggs, for the show car. It was designed by John Tjaarda, who was trained in aerodynamics and was influenced by Jaray. Tjaarda conceived the concept as a rear-engine vehicle based on work he had done before joining Briggs. This vehicle would provide the visual basis for the 1936 Lincoln-Zephyr. It benefited from a calculated and successful reworking by Ford's new design head, Eugene T. (Bob) Gregorie, who was surely influenced by the mixed reception accorded to the 1934 Chrysler Airflow.

If the Lincoln-Zephyr was considered by some to be America's first successful streamlined car, then the Airflow should be considered America's first aerodynamically designed production car. Chrysler went to the expense of investing in its own wind tunnel and even hired aircraft pioneer Orville Wright to assist in its development. Despite much being written about the public's rejection of the Airflow because of its unconventional styling (there were other considerations as well, such as poor initial quality), the car's general look was repeated by 1935's Peugeot 402 and the Volvo PV36. Both cars bear an uncanny similarity to the Airflow, and notably, the Peugeot achieved sales success.

Another significant aerodynamic vehicle was introduced in 1934, the Czechoslovakian Tatra T77. Designed under the leadership of Hans Ledwinka, this highly advanced vehicle incorporated a rear-mounted engine (like the Briggs/ Tjaarda Dream Car, William Bushnell Stout's Scarab, and Ferdinand Porsche's designs for what would become the Volkswagen). The Tatra went a step further, incorporating an air-cooled V-8 engine. The Tatra was also the only production car that licensed the Jaray patents (Jaray would latter seek compensation for patent infringement from both Chrysler and Peugeot). The Tatra would be produced in various forms, with continuous refinement, for 40 years. While the Chrysler Airflow and the Tatra T77 are not considered the most beautiful streamlined vehicles of the era, it should be remembered that their goal was achieving efficiencies *through* aerodynamics, and not just pure aesthetics. To put it another way, aerodynamic cars of the era were also considered streamlined, but not all streamlined vehicles were truly aerodynamic.

Fig. 6: The ferry boat *Kalakala* on Puget Sound, Washington, in the mid- to late 1930s. Photo by Underwood Archives/Getty Images

Fig. 7: 1933 Pierce-Arrow Silver Arrow. Photo Peter Harholdt

Custom-Bodied Art

The builders of custom coachwork would produce some of the most exquisite examples of streamline design during the 1930s. This was partially driven by the fact that they were less encumbered by cost constraints and timing and production realities faced by large-volume manufacturers. Additionally, they did not have to appeal to mass audiences. Rather, generally, just one well-heeled, fashionable, and tasteful client needed to be satisfied. By the latter half of the decade, many of these coachbuilders, enabled by some of the greatest designers of the period, would produce much of their best work and, arguably, the most highly regarded and sought-after vehicles of the classic era.

If streamline design was indeed an international trend, few embraced it like the French coachbuilders. Firms such as Figoni et Falaschi, Pourtout, Letourneur et Marchand, and Saoutchik, among others, would create dramatically streamlined bodies mounted on the expensive and sophisticated chassis of French luxury firms such as Delage, Delahaye, Talbot-Lago, and Bugatti, as well as other non-French marques. Some firms would design and build entire cars on their own. Gabriel Voisin, originally an aircraft designer and producer, would go on to design and manufacture some of the most distinct examples of streamlined coachwork, even among the other French luxury makes. The 1935 Voisin C28 Aérosport for example, featured a slab-sided, full-envelope body, which was very advanced despite retaining an upright vertical grille.

Bugatti, under the design direction of Jean Bugatti, would create many of its own bodies in-house. The 1936 Type 57S Atlantic is a brilliant example of a design that used a clever fabrication scheme—lightweight panels held together using a central spine and rivets—to create a signature design element. Georges Paulin, who would become the

(Left, top) Fig. 8: Coca-Cola Bottling Company, 1334 South Central Avenue, Los Angeles. Photo courtesy Library of Congress

(Left, bottom) Fig. 9: Aquatic Park Bathhouse/San Francisco Maritime National Historical Park Maritime Museum, 900 Beach Street, San Francisco. Photo courtesy Library of Congress

main designer for Pourtout, would pen designs for the Talbot-Lago T-150-C-SS, the Peugeot Darl'Mat, and the impressive 1938 Bentley Embiricos. Paulin's designs had a functional elegance to them. Author Richard Adatto states in his book *From Passion to Perfection*, which chronicles the French streamlined cars, "As always, his focus was first the aerodynamic performance of the body, followed by the beauty of the line."

But the most flamboyant of the French streamlined designs were surely done by Giuseppe (Joseph) Figoni. Inspired by aircraft design, Figoni would experiment with various interpretations of form development, with an emphasis on the ovoid. He would also do elaborate fender executions, going so far as to enclose both front and rear wheels (Fig. 10). Some of the designs could be questioned for their extravagant taste; Figoni was not known for his restraint with brightwork. But some designs, like the 1939 Delahaye 165M, created for the 1939 New York World's Fair, and his stunning series of Talbot-Lago "teardrop" coupes, are simply iconic.

Along with the French coachbuilders, Italian firms such as Carrozzerias Touring and Pinin Farina would also be highly influenced by the streamline trend. Touring would build some of its most celebrated designs on the Alfa Romeo 8C 2900B chassis. In 1937 at the Milan Motor show, Touring showed an open roadster and closed coupe based on this chassis, making them essentially racing cars in disguise. Enabled by exaggerated proportions and with exceptionally long hoods and beautifully realized surfaces, these vehicles were equally dramatic, purposeful, and elegant.

In Germany, Mercedes-Benz's Sindelfingen subsidiary would also build a number of streamlined designs in limited numbers on a variety of Mercedes chassis. Of these, the 1938 540K Autobahn-Kurier, while projecting an image very much in keeping with its name, would also use many of the design cues of the streamline vocabulary: large pontoon fenders, a fast rear profile, angled windshield, and skirted wheels. Built in very limited numbers, the 540K would retain a traditional radiator shell and headlamp execution, while the Autobahn-Kurier-derived Stromlinienwagen made additional strides toward effective streamlining (Fig. 11).

Other makes of the same period—like Chrysler, Peugeot, and Tatra—would strive to incorporate similar elements into their front surfaces.

Today, automobile manufacturers automatically include aerodynamic development as a standard step in the vehicle development process. Given the difficult requirements that manufacturers have to meet, a vehicle's aerodynamic performance is every bit as critical as its mass or engine size in balancing conflicting priorities, in an attempt to gain the greatest efficiencies. It's no surprise then that today's production cars are the most aerodynamically efficient ever, though you may not realize this from their appearance. Today's Audi A4, for example, a handsome, if conservative, upright sedan, has one of the lowest drag coefficients of any car on the road (Fig. 12). Yet there is little in its appearance that would suggest this.

What is memorable about the streamline era is that these vehicles visually celebrated the fact that they were dynamic objects. But more than that, they were also swoopy, elegant, sporty, sometimes flamboyant and even romantic, and very, very emotional; they were engaging and seductive. Even if most of these cars and motorcycles were not truly aerodynamic as we understand the science today, these historic vehicles are prized and admired because they are such strong visual statements. During an era that produced some of the finest classic vehicles of all time, the streamlined designs are rolling testimonials to a period when passion, artistry, craftsmanship, and technology truly came together.

Fig. 10: Madame Jeanne Falaschi alongside Delahaye, 1930s. Courtesy of Adatto Archive

Fig. 11: 1938 line drawing of Mercedes-Benz Stromlinienwagen. Photo courtesy Daimler AG

NOTES

1   Train Lover, "Streamlined Steam: Streamliner Memories, Memorabilia From The Silver Age Of Passenger Trains," posted August 17, 2012, www.streamlinermemories.info

2   Jeffery L. Meikle, *Twentieth Century Limited, Industrial Design in America 1925-1939*," 1979, Temple University Press, pp. 151.

Fig. 12: 2017 Audi A4 sedan. Photo courtesy Audi of America

THE SHAPE OF SPEED:
STREAMLINED AUTOMOBILES AND MOTORCYCLES, 1930 TO 1942

By Ken Gross, Guest Curator

The concept of streamlining has long been a fascination. Even before World War II, the confluence of aircraft design, the sleek shapes of railroad locomotives, the windswept elegance of fast speedboats, advanced highways like the German Autobahns, the Italian Autostradas, and our own high-speed Pennsylvania Turnpike, along with seminal, epoch-changing events like the 1939 New York City World's Fair, encouraged farsighted automotive designers to style truly streamlined cars that were functionally aerodynamic, faster, and more fuel-efficient.

"The Allure of the Automobile," which was presented at the Portland Art Museum in summer 2011, (Fig. 1) enthralled visitors with its lavish presentation of fine automobiles as beautiful kinetic art—"rolling sculpture," one could even say. The popularity of that initial presentation has encouraged the development of another such exhibition.

"The Shape of Speed: Streamlined Automobiles and Motorcycles, 1930 to 1942," presents a select grouping of rare and historic automobiles and motorcycles that demonstrates how auto designers worldwide translated the concept of increased aerodynamic efficiency into exciting machines that, in many cases, looked as though they were moving while at rest.

It should be remembered that early automobiles were closely related to the primitive horse-drawn carriages they were replacing. Automobiles were even popularly called "horseless carriages." Their engines were positioned under their occupants, their designs were generally square-shaped and upright, and little thought was given to any sort of flowing form. These pioneer vehicles were, fundamentally, designed exclusively around function. As the automobile evolved, that approach began to change. But it would take several decades before the automobile began to find its true form, and the influences for that evolution would come from many sources.

In the post-World War II period, advanced-design aircraft and rockets greatly impacted car designs. The results, translated into metal, were often fanciful and impractical. Think of all those towering fins ...

But from 1930 to 1942, automobile designs began to be more organic, emulating the classic teardrop shape which was thought, at that time, to be perfect for cheating the wind. The results were brought to life in automobiles and in a few motorcycles with then-startling shapes that looked as though they were ready to be embraced and caressed. Even if they weren't noticeably faster than their predecessors, they *looked* fast.

Not surprisingly, the conservative public balked in many cases. Sales of well-known brands slipped but then recovered as consumers tentatively embraced and accepted this brave new look. As we'll see, Lincoln waited, and when it introduced its graceful Zephyr, it rode an emerging wave of public understanding and gradual acceptance of the streamlined phenomenon to sales success.

The popular school of architecture called *Streamline Moderne* greatly influenced automobile styling in this period. It also had its effect on the shapes of radios, home appliances, transport trucks, and railroad locomotives, along with such disparate items as table flatware, water pitchers, toasters, cocktail shakers, and even pencil sharpeners. In the depths of the Great Depression, streamlining appeared as a positive, progressive, and even hopeful sign that bleak conditions would improve.

In his excellent introductory essay in this volume, David Rand discusses the origins of streamlining as a widespread design phenomenon, so I won't repeat his observations. My purpose is to discuss streamlining as it relates specifically to the automobiles and motorcycles that are the subjects of our exhibition.

But I do endorse the view of Barrie Down, who wrote, "Streamlining is an overused word that has two applications to automotive design and has often been applied incorrectly ... the first and true technical use of the term 'streamline' was first applied in the early days of the flying machine to describe the smooth flow of air over an aerofoil surface

Fig. 1: "The Allure of the Automobile: Driving in Style, 1930–1965," installation view, Portland Art Museum, Oregon. Photo by Robert Di Franco

that created lift … it was therefore easier to push a smooth or parallel shape through the air."[1]

"There were many designers who understood the benefits of streamlining." Down added. "They sought to improve the performance of their products by using aerodynamic theory."[2]

Interestingly, because streamlining wasn't a requirement for basic motoring—not yet at least—some of the aesthetics of these early efforts weren't too pleasing. In this exhibition, we have endeavored to choose attractive vehicles that successfully melded style and streamlining.

It's important to remember that automotive streamlining, when it first appeared, was equated with modernity as well as with efficient aerodynamics. The automobile, a child of the 20th century that was rapidly changing and evolving mechanically during the 1930s, was the perfect metal canvas

for streamlined design. Mark McCourt wrote, "… it was an era of unbridled, machine-driven technical advancement, of optimism in the unlimited possibility that the future held."[3]

"Automobiles reflected all this potential," McCourt continued. Despite the Great Depression, what he called "the vibrant promise of modernity and speed," was reflected not just in luxury goods, but in mass-produced items. "Some brave and revolutionary cars paid tribute to the *zeitgeist* with their overall design concepts, but most showed their Machine Age influence in small ways, in subtle and glorious details."[4]

The seventeen automobiles and two motorcycles selected for "The Shape of Speed," were carefully chosen to reflect this period's views and design practices. Streamlining was all the rage and its influence soon became widespread. Acclaimed architects and industrial designers such as Frank

Lloyd Wright, Norman Bel Geddes, Walter Dorwin Teague, Raymond Loewy, and Walter Gropius fell under its spell, as did many noted automotive stylists, race car innovators, and engineers like Jean Bugatti, Amos Northup, Joseph Figoni, Hans Ledwinka, John Tjaarda, Phillip O. Wright, Harry Arminius Miller, Mario Revelli di Beaumont, E.T. "Bob" Gregorie, Alex Tremulis, Harley Earl, Bill Mitchell, Gordon Miller Buehrig, and others.

The individual chapters in this catalogue will discuss in more detail how the chosen vehicles epitomize the streamlined style. For the purpose of this essay, we'll simply showcase some of the best examples and leave it to our readers and visitors to discover the myriad streamline influences that characterize and render special the machines on display.

Many of the early attempts to use streamlining in industrial design (aside from legitimate experiments by aerodynamicists like Paul Jaray, Richard Buckminster Fuller, and Wunibald Kamm) had, in Michael Lamm's words, "… nothing to do with the actual science of aerodynamics. Henry Dreyfuss called it 'cleanlining,' and … the principles of smoothing, integrating, unifying, and decorating weren't confined to automobiles." Radios, kitchen appliances, vacuum cleaners, even houses, "… were given great dollops of streamlining … it turned out to be a highly salable commodity."[5]

But not in every case: Chrysler's top engineers, led by the innovative Carl Breer, attempted to change popular automotive perceptions with a decidedly different approach to streamlining and design. The Chrysler Airflow emulated passenger trains, like the fabled Union Pacific *M-10000* and *Burlington Zephyr*.[6] But the fickle public, who had accepted the shift from classic steam locomotives to modern diesels, balked at buying a new car whose lines departed so abruptly from contemporary styling practice. Despite the fact that the Airflow's unitized body, smooth ride, many built-in safety elements, and flow-through cabin ventilation were superior to rivals, acceptance was less than anticipated.

When sales nosedived, Chrysler were forced to redesign the Airflow's pioneering shape and graft a conventional visage over its old grille (Fig. 2 and 3). Interestingly, the 1935 Airflow's front end and the new grille facelift were penned

Fig. 2 & Fig. 3: 1934 Chrysler Airflow grille, 1935 Chrysler Airflow grille. Photos by Peter Harholdt

by famed industrial designer Norman Bel Geddes. Chrysler's companion Airstream models were more conventional and they sold quite well. Our handsome 1934 Chrysler Imperial Airflow Model CV coupe evidences the original 'flat-nose' design, considered to be the purest example of the Airflow styling.

Emerging at about the same time, sleek, streamlined cars from Czechoslovakia's Tatra represented a similar approach. Named for the nearby Tatra mountain range, Tatra was the third oldest carmaker in the world, behind Mercedes-Benz and Peugeot, until 1999 when it switched to just building trucks. Tatra's engineers were Hans Ledwinka, who had his own unique approach to nearly every challenge, and Edmund Rumpler, a pioneer designer of aerodynamic cars. Protected by high tariffs in Czechoslovakia, Ledwinka began building dramatic-looking Tatra sedans with air-cooled, rear-mounted, V-8 engines.

In 1934, Tatra obtained a license from noted aerodynamicist and Graf Zeppelin designer Paul Jaray to build the Type 77, a full-sized fastback and the world's first truly aerodynamic car (Fig. 2). Wind tunnel tests of a scale model showed its coefficient of drag was 0.24. The actual car was 0.36 Cd, well below the 0.54 Cd for most cars of that period. The resemblance of the Volkswagen Beetle to the V570 Tatra that preceded it sparked a lawsuit by Ledwinka against Volkswagen and its designer, Ferdinand Porsche. When Germany invaded Czechoslovakia in 1938, the suit was dropped. After the war, the lawsuit was revived and VW had to pay a settlement of 3 million Deutsche Marks. Tatra's second large car iteration, the T77a, which we display in this exhibition, had an aerodynamic body, a central structural-tunnel floorpan, rear-wheel-drive, seating for five, and a front luggage compartment.

The Cord, named for company founder and famed industrialist Errett Lobban Cord, was built in Auburn, Indiana from 1929 through 1937. The earlier front-wheel-drive Model L-29 Cord was a highly advanced car in its day. Its successor, the stunning Cord Model 810/812, introduced in 1935/1936, embodied every conceivable Art Deco and streamlined design element, both inside and out. Designed by Gordon Miller Buehrig, who also penned the

acclaimed Auburn 851 Speedster, it's a perfect amalgam of stark angularity and luscious curves. The external flex-pipe exhausts on the Model 812 SC, displayed in this exhibition, became a hallmark of supercharged Auburn, Cord, and Duesenberg cars. Noted Cord owners included Olympic champion skater Sonja Henie, famed aviatrix Amelia Earhart, and orchestra leader Paul Whiteman. Compared with cars of its era, the Airflow Chryslers excepted, the curvaceous Cord epitomized streamlined styling at its best.

In mid-1930s France, Jean Bugatti was the talented 25-year-old son of automaker Ettore Bugatti, whose father in turn had been a designer of collectable, museum-quality furniture. Ettore's younger brother Rembrandt, aptly and challengingly named, was a superb sculptor of wild animals. Ettore's road-going and racing cars demonstrated a delightful flair that hinted at Art Deco influence, but it was Jean's skillful and repeated applications of dramatically curved forms, punctuated with controlled edginess and bold

Fig. 4: Tatra T77A. Photo by Peter Harholdt

streamlining, that created some of the most memorable automotive shapes of the pre-World War II period.

Jean's Bugatti's sexy Type 57 Aérolithe coupe design, which we have on display, employs a continuous line that races rearward from the handsome Bugatti horseshoe-shaped radiator (Ettore Bugatti, company founder and Jean's father, was a noted equestrian), rises in a picture-perfect arc, thanks to a raked windscreen, and tapers to a beautifully curved decklid. It's an eye-pleasing form that makes you smile at its audacity. The Aérolithe's body was constructed of magnesium, which was nearly impossible to weld without bursting into flame, so rivets were necessary to fasten the roof, body panels, and fenders together. A futuristic Jules Verne-esque effect was the pleasing result. Auto stylists today are still influenced by this Type 57 Bugatti's exquisite shape. Examine this car closely and be prepared to be delighted.

We featured a Pierce-Arrow Silver Arrow in the "Allure of the Automobile" exhibition. An important contributor to the advance of streamlining in the mid-1930s, it is important to cite here, even though it's not in this exhibition. Nearly out of business as the Depression endured, Pierce-Arrow was a proud, Buffalo, New York-built luxury make that vied with Packard and Cadillac. Pierce-Arrow cleanly won a competition at the 1933-34 Chicago Century of Progress exhibition with its shapely Silver Arrow, a streamlined leviathan limousine armed with a locomotive-like 12-cylinder engine and replete with deliciously Art Deco elements, from its sweeping flush fenders that hid the show car's side-mounted spares, to "... an intriguingly sinister rear window treatment (that) perched two slivers of glass in a periscope-like rear peak." It ensured that rearward vision was terrible, but it looked great. Art Deco accents and details made the Pierce appear as if Erté himself had had a hand in its conception. Phillip O. Wright, a prolific auto stylist,

was largely responsible for the design. Just five Silver Arrows were sold, at a then-heady $10,000 each. Pierce-Arrow and Studebaker's attempts to market a watered-down version of Wright's design failed, as did Pierce-Arrow, in 1938.[7]

William Bushnell Stout, a pioneering and successful aircraft engineer, designed the famous Ford Tri-Motor all-metal airplane. Stout's then-radical Scarab pays homage to the school of Art Deco design, and it's beautifully streamlined as well. Oddly beetle-shaped, aircraft-inspired, and a forerunner of the minivan, the Scarab's fluted headlamps, Egyptian-style amulet nose badge, pert rows of 'whiskers' in lieu of a conventional grille (its Ford flathead V-8 was rear-mounted), fan-shaped cooling trim, and the prominent peak that divides the hood and the front windows are simply delightful styling conceits.

Only a handful of Stout Scarabs were ever made. The lethal combination of high price, high risk for customers in buying an

Fig 5: Voisin C25 Aerodyne. Photo by Michael Furman

Fig. 6: Mercedes-Benz Stromlinienwagen on the road. Photo courtesy Daimler AG

unknown make, and the Stout's sheer unconventionality made certain it became only a lasting footnote to the streamlined influence of its era. Ironically, when people see the Stout on display today, knowing nothing of the car at first, they immediately grasp its significance, because they have the modern minivan as a frame of reference. Lacking that in 1935, the Scarab, very expensive for its day at $5,000, merely appeared to be a pricey curiosity. But streamlined cars didn't have to be expensive. The Steyr Model 50 "Baby" was an affordable European small car, built to do battle with similarly-conceived tiny tots from Volkswagen, Tatra and others.

The worldwide acclaim for Charles Lindbergh's era-inspiring solo flight across the Atlantic to Paris in 1927 helped to focus many auto designers on the merits of aerodynamics and aircraft shapes. At first it was simply the look that captivated these designers. Later, clever engineers and some aerodynamicists began to realize that there was an efficiency component to streamlining. Their next task was to ensure that, while making cars that resembled wingless airplanes, the designs themselves remained attractive and acceptably car-like, without alienating buyers who were used to more upright styling interpretations.

There was a worldwide interest in aircraft *technology* (as well in air travel) and several aeronautical engineers examined

Fig. 7: Princess Stella de Kapurthala with her Talbot-Lago Teardrop Coupe in Paris, 1938. Courtesy of Figoni Archive/Adatto Archive

a possible transfer of technology from aircraft to automobiles. Gabriel Voisin, who pioneered aircraft production in France and made his fortune as a supplier during World War I, was one of those. Voisin used airplane construction techniques and materials, and he cared not a whit if people liked his cars or they didn't (Fig. 5). Paul Lewis wanted to capitalize on aircraft knowledge and construction. His Airomobile looks like nothing so much as a small plane that lost its wings wandering on the highway. Despite Lewis's having traveled 45,000 miles around the country demonstrating the merits of the Airomobile's streamlined design, investors failed to materialize.

Some designers only got streamlining "half right." In England in the 1920s and 1930s, what became known as the "Airline" design practice popularly retained recognizable radiators, grilles, and front-end elements of marques like Rolls-Royce, Daimler, Bentley, Jaguar, and even Singer and MG, while the designers streamlined, rounded and tapered the rear aspects of the cars. The front-end visages remained largely the same, but from the rear, these free-flowing bodies appeared to be quite modern. Mercedes-Benz offered the rare Autobahn-Kurier in this idiom, retaining the standard Mercedes-Benz 540K's haughty grille with all of its baroque opulence. The one-off Mercedes-Benz Stromlinienwagen, on display in this exhibition, has a streamlined grille, possibly the only such treatment on a big, road-going Mercedes-Benz of this era. From the windscreen to the rear, the body is curved and tapered (Fig. 6).

Several of the automobiles in this streamlined collection were one-of-a-kind when they were built. The Bendix SWC prototype sedan was commissioned by the famous Detroit-based Fisher brothers, whose huge Fisher Body subsidiary built car bodies for General Motors. Engineered by Roscoe "Rod" Hoffman, a prolific inventor, and fitted with an unusual X-8 engine in the rear, the unitized, streamlined, and well-built Hoffman incorporates many clever ideas. But it never went into production.

Edsel Ford, president of Ford Motor Company at age 25 and the only son of auto magnate Henry Ford, was the consummate car enthusiast. In charge of the Lincoln brand, he sought a way to stop staid old Lincoln's flagging sales.

Fig. 8: 1942 Alfa Romeo 6C 2500 SS Bertone. Courtesy Archivio Collezione Lopresto

Edsel became interested in a rear-engine car design done for the Briggs Body Company by futurist John Tjaarda. This prototype, called the Sterkenburg, became the basis for the Lincoln-Zephyr, a stylish 12-cylinder streamlined model that sold for less than half the price of the big Lincolns. After Chrysler's stumble, Edsel might have been wary of offering a streamlined car, but the Zephyr was profoundly beautiful *and* streamlined, and the market loved it. The Lincoln-Zephyr chassis became the basis for the later, landmark Continental.

Parisian carrossiers Claude Figoni and Ovidio Falaschi were, by practice and as expressed in their own advertisements, the "Couturiers of the Automobile." Their *ateliers* hand-crafted gorgeous automobiles that truly resembled Paris gowns on wheels. Voluptuous shapes, fully skirted front fenders, dramatic speedline themes repeated in several places, low windscreens, chopped rooflines, long hoods, selfish cabins, teardrop motifs ... all contributed to the unique appearance of a Figoni and Falaschi road-going dream car. French fashion enthusiasts pioneered the creation of the

Concours d'Elegance, in essence a closely judged fashion show for fine automobiles, each one accompanied and accented by a beautifully dressed woman whose designer ensemble was color-coordinated and themed to complement the car itself (Fig. 7).

Figoni-built classics dominated these Concours exhibitions with their audacious, often impractical, but memorable designs, usually built on expensive sporting chassis from top-line marques like Delahaye, Bugatti, Delage, and Talbot-Lago. The underpinnings of these stunning, streamlined cars were often full competition models—their sleek aluminum skins were exquisitely molded, painted in bright hues or in glittering black, then accentuated with delicate kisses of polished alloy and discreet chrome trim.

In the 1930s, it was commonly believed that the tear drop was the perfect aerodynamic shape, so it became an

Fig. 9 and 10: 1941 Chrysler Thunderbolt Roadster. Courtesy of FCA North America

oft-repeated theme in streamlining. Our T-150C-SS Talbot-Lago Teardrop Coupe is one of a limited series of expensive custom coachwork exercises on high-performance chassis. Wealthy buyers could specify individual touches, such as faired-in or free-standing headlamps, sunroofs, long or short tails, vestigial fins, hood louvers, and all manner of fancy interior fabrics and special hides. As if limited-production custom cars weren't unique enough, the carrossiers would entertain client suggestions of color selections, accessories, fitted luggage, disc or wire wheels, standard or pre-selector gearboxes, and other personalized choices, adding up to a bespoke car.

At the end of the 1930s two cars, in two different ways, took streamlining to new heights. The dramatically different Sharknose Graham, created by the inimitable Amos Northup, was a late-decade attempt to save a faltering old company. These striking cars attracted some attention—and they are wild-looking by today's standards—but alas, they did not provide the sales burst needed for Graham to survive. The Alfa Romeo 6C 2500 sports coupe in our exhibition, designed by the incomparable Mario Revelli di Beaumont, and credited to Carrozzeria Bertone, was built on a modified early Alfa chassis and constructed *during* the war years (Fig. 8). Its streamlined appearance and splendid fadeaway fenders undoubtedly influenced the postwar Bentley Continental.

Motorcycle designs in the 1930s did not typically have streamlined bodywork. The two machines on display represent rare approaches to aerodynamics for two-wheelers. The BMW R7, a long-lost, recently restored concept bike, has fully streamlined bodywork and artfully shaped fenders. Built in 1934, it was probably thought to be too expensive an approach. While a few of its design elements appeared on standard BMW motorcycles, the R7 did not go into series production. O. Ray Courtney's Henderson KJ was one man's interpretation of the "motorcycle of tomorrow." He was willing to build copies for clients, but in Depression-wracked 1936 there were no takers for this streamlined machine, or for a subsequent fully-enclosed and streamlined postwar two-seater design that Courtney developed.

The Chrysler Thunderbolt, created and shown just before World War II began, is both a continuation of '30s-era Art Deco and streamlined themes seen in the Scarab, for example, and an insightful prediction of where cars would be heading in the postwar period: slab sides, a pancake hood, hidden headlights, minimal decoration, horizontal trim elements evoking speed and progress—all these signaled the car of tomorrow.[8] Sadly the Thunderbolt, and its companion four-door phaeton called the Newport, were limited in production, partly by cost and because the American auto industry would soon be gearing up to become "The Arsenal of Democracy." (Fig. 9 and 10)

Automobiles, by the time the period between the two World Wars sadly ended, had in Mark McCourt's well-chosen words, "… transcended pure function and become the ultimate personal accessories, desirable for their design as much as for their speed, luxury and practicality … (speaking) volumes about their owners' tastes and their parent company's engineering bravado. And when automobile production resumed after World War II, the world was in a different place, with a new sense of the awesome (as well as awful) power of machines; automobile design would soon take a different path, leaving the cars and cues of this late interwar period unique and forever celebrated."[9]

"The Shape of Speed" exhibition proudly presents the best of the 1930-to-1942 decade, giving us a close look at cars and motorcycles that define the genre. We're greatly indebted to the generous owners who have had the foresight to collect and preserve these wonderful machines, and we thank them for sharing their treasures with an appreciative audience. While these cars are perfectly pictured herein, and they have been artistically displayed and spotlighted in the galleries of Portland Art Museum, it's important to remember they are rolling sculpture, really kinetic art, and capable of dynamic function. Just imagine any one of these fabulous cars coming to life …

The late Ken W. Purdy wrote it best. Calling Bugatti, one of our honored marques, "… the Prince of Motors," Purdy challenged his readers to "… imagine a string-straight, poplar-lined Route Nationale in France on a summer's day. That growing dot in the middle distance is a sky-blue Bugatti coupe rasping down from Paris to Nice at 110 miles an hour …"[10]

NOTES

1   Barrie Down, *Art Deco and British Car Design: The Airline Cars of the 1930s*, Veloce Publishing, Ltd., Poundbury, Dorchester, Dorset, England, 2010, pp. 14.

2   Down, ibid., pp. 14.

3   "Art Deco and the Automobile," by Mark J. McCourt, *Hemmings Classic Car*, December, 2012, pp. 22.

4.   McCourt, ibid., pp 22-25.

5   Michael Lamm and Dave Holls, *A Century of Automotive Style*, Lamm-Morada Publishing Company, Inc., Stockton, CA, 1996, pp. 79.

6   Charlotte Benton, Tim Benton, and Ghislaine Wood, *Art Deco, 1910-1939*, Bulfinch Press, Boston, New York, London, 2003, pp. 365.

7   Lamm and Holls, ibid., pp. 231.

8   Jonathan Stein, Editor, *Curves of Steel: Streamlined Automobile Design at the Phoenix Art Museum*, Coachbuilt Press, Philadelphia, 2007, pp. 58-61.

9   McCourt, ibid., pp. 29.

10   Ken W. Purdy, *The Kings of the Road*, Atlantic: Little, Brown & Company, Boston, 1949, pp. 5.

OTHER REFERENCES:

Donald J. Bush, *The Streamlined Decade*, George Brazilier, New York, NY, 1975.

Michael Zumbrunn and Robert Cumberford, *Auto Legends: Classics of Style and Design*, Merrell Publishers, London and New York, 2004.

"Streamliners: Sleek, Stylish and Stable at Speed," by Charles Dressing, Amelia Island Concours d'Elegance Program, pp. 124-131.

PART II:
THE AUTOMOBILES AND MOTORCYCLES

1930 HENDERSON KJ STREAMLINE

COLLECTION OF FRANK WESTFALL, NER-A-CAR MUSEUM

Harley-Davidson in Milwaukee remains the only major surviving American motorcycle company, but decades ago there were several rival firms whose superb products challenged Harley in every respect. Indian motorcycles were built in Springfield, Massachusetts, until the early 1950s; Henderson and Excelsior, manufactured by the Excelsior Motor Mfg. & Supply Co., were technically competent machines. Just before the stock market crash of 1929, Indian Motorcycles, Henderson and Excelsior, built in Chicago, and the venerable Ace 4, built in Philadelphia, competed during what T. A. Hodgdon called "The Golden Age of the Fours."[1]

The 1929 Henderson Model KJ, better known as the "Streamline" model, was capable of an honest 100 mph, thanks to its 1,200-cc, 40-bhp, inline four-cylinder engine; a sturdy, five-main-bearing crankshaft; and a downdraft carburetor. Suspension advances like leading-link front forks and an illuminated speedometer that was built into the fuel tank helped make the "Hendee" the choice of countless municipal police departments.

Motorcycle customization in the 1930s commonly consisted of stripped-down machines called "bobbers" or "bobtails." Nonessential parts were eliminated; the front fender was usually discarded and the rear fender was trimmed (or "bobbed"). The result resembled a hill-climb racer. O. Ray Courtney took the opposite approach: in an era when streamlining applied, if at all, only to the shape of a bike's fuel tank and fenders, and when enclosed bodywork was virtually unknown on production two-wheelers except for a few racing machines, Courtney fabricated beautifully formed bodywork for his four-cylinder KJ.

According to vintage motorcycle authority Ed Youngblood, Courtney, who was born in New Cardon, Indiana, in 1895, worked for Central Manufacturing Company, a supplier of body panels to Auburn and Duesenberg just after World War I. An active motorcyclist, Courtney believed that the motorcycle industry erred on the side of power and speed, and failed to provide weather protection and luxury for its riders.[2]

In 1934, starting with a 1930 Henderson Model KJ four-cylinder motorcycle as a platform, Courtney devised and built a radically streamlined body shell that was unlike anything ever done on two wheels. Courtney's sleek one-off had a curved, vertical-bar grille, reminiscent of the just-introduced Chrysler Airflow, and the rear resembled an Auburn boat-tail speedster. The bike's modified, 10-inch wheels were considerably smaller and wider than the original Henderson wheels. The Streamline's envelope body panels were hand-formed of steel with a power hammer. Courtney modified the Henderson's telescopic front fork, and the rear suspension was built from automotive parts.[3]

Youngblood commented that "...the streamlined Henderson was a pure concept vehicle, built to express modern concepts and Courtney's artistic vision. The conservative motorcycle community of the era did not understand it. References in print used the term "Buck Rogers," treating it like something out of a futuristic cartoon.'[4]

The Courtney Streamline was stunningly beautiful but impractical. A tall person would find the seating position cramped. The complex-curved, all-steel body would have been heavy and difficult to make. In 1941, Courtney filed a patent for a motorcycle design with fully enclosed fenders. Perhaps he was influenced by the fact that Indian had introduced its partially skirted fenders in 1940, and that motorcyclists were becoming more accepting of this trend.

Courtney never tried to market his enclosed-fender design. He went on to work for Oldsmobile, and later Kaiser-Frazer.

At the same time, he and his son, Ray William Courtney, established a company to build and repair sheetmetal panels for racing cars. In 1950, Courtney built a prototype of another streamlined motorcycle—this time a two-passenger machine—that he called the Enterprise. As Henderson had long been out of business, the Enterprise prototype was powered by an Indian Scout V-twin engine. Courtney's radical bike appeared on the cover of *Popular Science*. He offered to build copies of the Enterprise for $2,500, if a customer supplied an engine. That never happened; the Enterprise was more expensive than a typical low-priced new car.

When Courtney passed away in 1982, the disassembled KJ Streamline and the Enterprise were both acquired by Ron Finch. He sold the two prototype bikes and a collection of historic development material about them to Mike Gagliotti, a collector from Syracuse, New York. Frank Westfall, the current owner, bought the disassembled KJ and the fully intact Enterprise from Gagliotti. The Henderson needed a complete, ground-up restoration, which was accomplished by another Syracuse-based contributor, Pat Murphy.

The restored Streamline debuted at the Antique Motorcycle Club of America National Meet in Syracuse, and caused a sensation when Frank Westfall rode it at the show. O. Ray Courtney's ideas were certainly ahead of their time, and the KJ Streamline and the Enterprise survive to prove his point.

—KNG

NOTES

1 T. A. Hodgdon, *Motorcyling's Golden Age of the Fours* (Lake Arrowhead, CA: Bagnall Publishing, 1973).

2 Ed Youngblood, writing on his www.motohistory.net website, and his article in www.antiquemotorcycle.org.

3 Youngblood, ibid.

4 Youngblood, ibid.

Henderson photos by Peter Harholdt

1934 BENDIX SWC SEDAN

COURTESY OF THE STUDEBAKER NATIONAL MUSEUM

Here's an advanced, one-of-a-kind streamlined car most people have never seen. Although it originally broke cover about the same time as the Chrysler Airflow, there was no connection between the two. Vincent Hugo Bendix had tried briefly to market a car under his own name, the Bendix Buggy, from 1907 to 1909. Then he struck it rich after he invented the Bendix starter drive—the spring-loaded pinion gear that connects an electric starter to an automobile's flywheel—and successfully sold it to Chevrolet. Bendix made even more money manufacturing and selling innovative hydraulic brake systems. He then expanded his automotive empire to more than 100 companies, including Bendix Aviation, Bendix Brakes, Stromberg carburetors, Scintilla-Vertex magnetos, and Pioneer instruments. When Bendix acquired the Bragg-Kliesrath company, makers of vacuum brake boosters, Victor Kliesrath, a racing enthusiast, became his vice president of engineering.

In 1931, Bendix acquired the defunct Peerless automobile plant in Cleveland. He longed to develop his own car, but the Depression was hardly an opportune time. Nevertheless, Kliesrath headed up a top-secret Bendix project to develop an advanced automobile. The work was kept undercover because Vincent Bendix couldn't risk angering his large Detroit automaker clientele. A former Packard engineer, Alfred M. Ney, was hired to design the car. To keep the project hidden, it was to be manufactured under a dummy corporation, the Detroit Steel Wheel Company. A small group of talented men worked in secret to build the prototype. The frame was of sturdy box-section steel construction with subframes fore and aft for the engine and rear suspension. Ex-Fisher body designer William F. Ortwig did the styling. Ortwig told historian Michael Lamm years ago that he

had no idea what the Chrysler Airflow would look like when he penned the SWC in 1932, so any resemblance is purely coincidental.[1]

Ortwig's advanced design for the SWC originally mandated a lightweight, unitized, streamlined body with a total car weight of 1,500 to 1,700 lbs. But Ney told Lamm, "Bendix and Kleisrath were in a hurry to build it, so they had Bill Ortwig design a conventional-for-the-time, wood-framed steel body with an aluminum hood, doors, and fenders."

By the time the SWC was completed, new DeSoto headlamps, grille, and bumpers were used, lending the impression that the design copied the DeSoto Airflow, but the car's shape had been determined much earlier. Built in Benton Harbor, Michigan, the project took two and one-half years to complete. The prototype was shipped to Europe, where the car was shown to several manufacturers including Bentley, Bugatti, and Renault, to encourage them to purchase Bendix components. There doesn't seem to have been any attempt to manufacture and sell it as a futuristic car.

The SWC's engine was a much-modified, rubber spool-mounted 169.5-cid L-head Continental six. The engine was reversed 180 degrees to mount a three-speed transaxle at its front. Fitted with a Bendix "Finger-Tip Control" preselector, this transaxle became the SWC's front-wheel-drive unit. Curiously, the I-6 engine was fitted with an aluminum head that incorporated a cast iron "steam dome" to help cool the engine via vaporization. There was no cooling fan. The front suspension was fully independent with rubber-mounted A-arms (which were later changed to quarter-elliptic leaf springs), stressed rubber bushings, and Weiss-jointed drive axles. Then-fashionably wide General Jumbo 14-inch tires, inflated to just 8 psi, also acted as suspension elements. The rear suspension was fully independent as well.[2]

In 1934, the sleek fastback Bendix SWC sedan cost some $84,000 to build, the equivalent of just over $2 million in today's currency. The prototype's hydraulic, four-wheel, duo-servo Bendix drum brakes, one of the car's many advanced features, were adjustable from the outside. The wheels themselves bolted to the periphery of the drums. Slotted wheel covers directed air into the brake drums. Inside the cabin, suspended pedals were a modern touch.

General Motors owned 25 percent of Bendix at that time. Hard hit by the Depression, Vincent Bendix's companies faltered and the stock plummeted nearly $100/share by 1933. GM brass disapproved of Vincent Bendix's lavish lifestyle and there was a fight for company control. By 1937, GM had forced Bendix out; he declared bankruptcy, lost his companies, and sadly died in 1945 at age 63. The SWC's fate was something of a mystery. It dropped out of sight and was hidden under a cover in a warehouse until it was re-discovered in 1967 by a Bendix employee named Gene Wadzinski. He managed to free the frozen engine and get the SWC running.[3]

Given the mixed reception afforded the Chrysler Airflow, the fact that Vincent Bendix was beholden to his many Detroit clients, and build compromises had made the car heavier than its designer intended, the SWC was arguably doomed from the outset as a possible production car. But looking back, its smartly aerodynamic styling and innovative drivetrain were advanced for the time.

—KNG

NOTES

1 "FWD Bendix SWC driveReport," by Michael Lamm, *Special-Interest Autos*, Nov/Dec 1971, pp. 40-45.

2 ibid., pp. 42.

3 ibid. pp 41.

Bendix photos by Peter Harholdt

1934 BMW R7 CONCEPT MOTORCYCLE

BMW CLASSIC COLLECTION

BMW is one of just three contemporary automobile companies that also produce motorcycles using the same brand name (the others are Honda and Suzuki). The Bayerische Motoren Werke AG (Bavarian Motor Works) started in 1916, as Rapp Motorenwerke GmbH in Munich, Germany, and merged with Bayerische Flugzeugwerke AG, an aircraft manufacturer (the blue-and-white BMW logo roundel represents a spinning propeller). Motorcycles, long an important part of BMW's business, helped the company regain its footing after both world wars.[1]

From the outset, BMWs were different from their rivals. Chief engineer Max Friz developed an air-cooled, horizontally opposed twin-cylinder, side-valve engine for a short-lived German make called Helios, then improved it substantially for the new BMW R32. Shorter, lower, and lighter than the Helios, the more advanced BMW R32 used a rigid tubular frame and a quarter-elliptic leaf-spring cantilever front suspension. Instead of a messy chain, it had a three-speed transmission, a flexible coupling, and shaft drive. The modern motorcycle was born. BMW enjoyed initial competition success, winning the International Six Days Trial in England, with an R37, in 1926. By 1929, BMW R11 and R16 models had pressed-steel frames and even more innovative styling. The 1928/9 R63 was a 750-cc overhead valve twin, and by 1929, the R16 Series III was fitted with dual carburetors.[2]

Designed by Alfred Böning, the startling BMW R7 was built as a one-of-a-kind concept bike in 1934. There was really nothing like it in the marketplace. The frame was constructed of pressed steel. The front and rear fenders were gracefully skirted; the fuel tank, along with streamlined and fluted partial side covers for the engine, was sensuously styled and beautifully integrated. The overall side elevation made the bike appear as though it were chiseled out of a solid block of silver and quartz.

The engine was an 800-cc opposed twin, designed by Leonhard Ischinger, with a forged one-piece crankshaft, one-piece individual cylinders, and finned cylinder heads. The camshaft was positioned under the crankshaft, so the heads could be located a little higher than on previous BMW flat twins, resulting in a more efficient valve location and more ground clearance than most motorcycles of that era. The exhaust pipes were exquisitely curved, and the shape of the mufflers beautifully complemented the flowing lines of the bodywork. Even the handlebars were different; they were stamped steel, and their aerodynamic shape undoubtedly helped with this machine's streamlining.

There was a four-speed transmission, and the gearshift lever was located, automobile style, in a slotted gate adjacent to the fuel tank, high on the right side of the motorcycle. There was no rear suspension; instead the rear seat was sprung, as was the case on so many motorcycles of that era. A hinged inspection plate on the right side of the engine cover opened to allow access to the upper part of the engine. The speedometer was a drum-type unit; an oil pressure gauge was located atop the fuel tank.

Finished in period BMW glossy black and white-striped livery, the flowing single-seater was very advanced for its era. It is believed to have been shown several times, and then it was lost and long forgotten. Reportedly, this stunning cycle was stored in a crate for nearly seventy years. It was discovered in 2005, and the restoration, by BMW in-house restorers Armin Frey and Hans Keckeisen, took two years. The R7 certainly influenced several succeeding BMW bike models, but none of them possess its unique and graceful blend of Art Deco styling, advanced engineering, and enclosed coachwork.[3]

In August 2012, the BMW R7 won first place in the BMW Motorcycle Class at the Pebble Beach Concours

The R7 certainly influenced several succeeding BMW bike models.

d'Elegance. Klaus Kutscher, from BMW Classic in Munich, brought the bike to Pebble. "They only made one of these," he told the *Tonight Show* host and motorcycle collector Jay Leno. "It would have been too expensive to produce," he opined, "so they didn't build it."[4] BMW was hardly a mass marketer of motorcycles—its bikes were expensive—but it is believed that Munich's management felt that there was no market for a truly luxurious machine in the depths of the Depression. Thankfully, this marvelous motorbike has survived as a lovely reminder of what might have been.

—KNG

NOTES

1   Halwart Schrader, *BMW: A History* (Princeton, NJ: Princeton Publishing, 1979), 14–19.

2   L. J. K. Setright, *The Story of BMW Motorcycles* (London: Transport Bookman Publications, 1977), 15–18.

3   "Rare BMW Stars at Pebble Beach," *Cycle World*, September 10, 2012, available online at www.cycleworld.com.

4   "BMW R7 at Pebble Beach," *Jay Leno's Garage*, www.jaylenosgarage.com.

BMW photos by Peter Harholdt

1934 CHRYSLER IMPERIAL MODEL CV AIRFLOW COUPE

COURTESY DAVID AND LISA HELMER

When the radical Airflow model was first released in 1934, the Chrysler Corporation was only 10 years old, having evolved from Maxwell. One of the young company's strengths was engineering excellence, an outgrowth of Walter P. Chrysler's early experience as a railroad master mechanic.[1] That also explained why Chryslers and DeSotos tended to be rather conventional.

All that changed with the introduction of the Airflow in 1934. The man behind this new direction was talented engineer Carl Breer, one of a trio of engineers known as "The Three Musketeers."[2] According to a story published in *Automotive News*,[3] Breer was inspired by seeing a formation of military airplanes. This experience caused him to consider wind resistance and air pressure, and led him to consult with aviation pioneer Orville Wright in building a wind tunnel.

Wind tunnel testing revealed that the typical, square-shaped 1930s automobile was more aerodynamically efficient when driven backwards, a revelation that led directly to the Airflow line. While Breer was focused on aerodynamic shapes, his colleague Fred Zeder designed a strong unibody structure. The resulting Trifon Special prototype shared many of the visual hallmarks that later appeared on production 1934 Airflows.

Snub-nosed, and with the passengers positioned in front of the rear axle, the bold new Airflow Chryslers and DeSotos lacked the long hoods and sweeping fenders of conventional rivals. A lavish chrome waterfall grille, side strakes, skirted fenders, multi-bar bumpers, and tubular chrome-plated seat frames clearly showed Art Deco influences, and demonstrated brilliant detailing. Unlike the more flamboyant French and European expressions of the Art Deco theme, these cars were more linear and streamlined, and literally engineered to cut through the air.

Although the new Airflow was miles ahead in safety and strength, its unconventional design was also miles ahead of public acceptance. Sales figures were disappointing. Chrysler models sold 11,016 units, while 13,900 examples of the DeSoto found buyers.

The lackluster sales created an imperative for a quick facelift for 1935. According to Karla Rosenbusch, writing in *Automobile Quarterly*, "Consultant designer Norman Bel Geddes was contracted to create a new look for both the Chrysler and DeSoto lines, in the hope of reviving moribund sales."[4]

Although the unibody remained unchanged, the ensuing facelift included an extended hood with a vee-shaped prow. The new nose treatment was more marketable to the car-buying public. The bumpers were simplified and now evidenced a subtle expression of Art Deco influences. Chrome accents were thoughtfully employed and the overall appearance was less radical. Arguably, the most successful application of the facelift was to the 1935 Chrysler. In a clever marketing ploy, owners of 1934 models could order the new body components to have their older Airflows updated.

Featuring the purest Airflow design, this 1934 Model CV was a larger version of the standard Airflow CU, on a longer, 128-inch wheelbase, suitable for five passengers. Its engine was a 325.5-cid straight-8 that developed 130 bhp.

The Model CV was equipped with vacuum-assisted four-wheel power brakes. Chrysler advertising proudly called it, "The Magic Carpet of 1934!"

Current owner and lender David Helmer, of Macungie, Pennsylvania, found this car literally in boxes in Chicago, fortunately with its unibody shell intact. He then spent ten years seeking and gathering up missing parts. The restoration work was performed by Airflow expert Gary Hoover of H & H Collision & Paint in Alliance, Ohio. Hoover has restored more than 30 Airflow cars. The few parts he couldn't find for this coupe, he skillfully fabricated.

The 1934 Chrysler Imperial Airflow CV is one of the most significant early streamlined American cars, and this rare coupe, resplendent in black, is one of only two examples known to survive, of just 212 built. Extensively shown since its completion in 2017, it has won awards at Amelia Island, Greenwich, The Elegance at Hershey, and the Concours of America at St. Johns. This striking coupe is a wonderful example of a brilliantly conceived and well-executed machine that was simply too modern for its time.

—JAS/KNG

NOTES

1   Karla A. Rosenbusch, "Walter P. Chrysler, an American Workman," *Automobile Quarterly*, Volume 32, Number 4, (1994) 6.

2   Karla A. Rosenbusch, "One for all and all for One! Chrysler's Three Engineers," *Automobile Quarterly*, Volume 37, Number 3 (1997) 70.

3   *Automotive News*, January 17, 1964

4   Karla A. Rosenbusch, "In the Eye of the Beholder, Chrysler's Airflow," *Automobile Quarterly*, Volume 38, Number 3 (1998) 66.

Chrysler Airflow photos by Peter Harholdt

1934 GRAHAM COMBINATION COUPE

COURTESY CHARLES MALLORY

Perhaps no other 1930s-era car epitomizes the American auto industry's interest in stylish streamlining more than the bold Graham "Spirit of Motion." Popularly called the "Sharknose Graham," this eye-catching design was the work of Amos E. Northup, the talented chief designer for the Murray Body Company. Northup's memorable work ranged from the pert, low-priced Willys Model 77 to the now-classic Reo Royale. He ventured into streamlining with the aptly named Graham Blue Streak in 1932. While not as dramatic as the bulbous Chrysler Airflow of two years later, the Blue Streak's flowing fenders and canted roofline were visibly more windswept than most of its upright competition.

The market reacted favorably at first to the Blue Streak. But despite offering a supercharged six-cylinder engine in its mid-priced field, Graham sales needed a late-1930s boost. Northup countered with this very forward-looking shape, calling it the "Spirit of Motion." Writing in the inaugural issue of *Special-Interest Autos*, editor Mike Lamm noted, "The striking, leaning-into-the-wind profile copied old racing photographs, in which the camera shutter made all the fast cars look like they were trying to outrun themselves. Even standing still, the Graham seemed to be moving."[1] *Hemmings Motor News* editor David LaChance added, "Designers of the day considered the Streamline Moderne look to be not just a fresh idea but a tonic for a Depression-weary nation."[2]

Henry Dreyfuss, who designed the *20th Century Limited* locomotive, which sped from New York to Los Angeles, commented, "Streamlining is the first new and uniquely American approach that the public could associate with progress and a better life." To be fair, Paul Jaray and other visionaries on the Continent were also offering sleek, streamlined shapes. Rail travelers could see progress daily in

the modernistic treatment of steam and diesel locomotives. Graham proudly showed the Spirit of Motion in many European auto salons, mostly to favorable acclaim. A few coachbuilders like Count Alexis de Sakhnovsky offered custom coachwork on the Graham chassis, retaining the distinctive slanted nose while reshaping the body and fenders.

But times were still tough, and the overall public response to Graham's windswept look was primarily one of rejection. Just 4,139 Sharknose Grahams found homes in 1939, nearly 10,000 fewer sales than the 1938 Grahams. Hindsight is 20/20, but it's probably safe to say that the Northup's severely raked grille design was a little too radical for its day—the same fate suffered by Chrysler's Airflow. Facing intense competition and without funding to finance a complete makeover, Graham's 1940 cars featured a more upright grille. Sadly, Amos Northup had suffered a fatal head injury in 1937, when he slipped on an icy walkway in front of his Detroit home. He didn't live to see his design's poor market acceptance.

Besides its unmistakable visage, the Graham Sharknose was quite a performance car for its day. The 217.8-cid flathead I-6, supplied by Continental, was equipped with an optional centrifugal supercharger, similar to the unit employed by Auburn, Cord, and Duesenberg. The water-cooled blower developed 4 psi and turned at a heady 24,000 rpm, six times crankshaft speed, while it developed 116 bhp (120 bhp in 1940) and 182 lb/ft of torque. "Duo-ratio" steering made parking easier and improved response at higher speeds. Hydraulic brakes were included. Warner gear overdrive was an $85 option. Anticipating the methods of today's automakers, several custom trim groups could be ordered. At $1,275 f.o.b. Detroit, the Graham was less than half the price of a Cord, with decent performance and a top speed of nearly 90 mph.

Styling details on the daring Sharknose still attract the

discerning eye. The windshield and side window glass are large and airy. On this stylish five-passenger Combination Coupe, thin B-pillars and window rails (a feature that would be shared by the new Mercury), contribute to a fresh and modern "hardtop" appearance. The Spirit of Motion's massive headlights are Art Deco-inspired, with elaborately scribed lenses and a square-ish shape that must have encouraged considerable comment in the day.

By early 1940, however, Graham was nearly out of time. The company bought body dies from the defunct Cord Corporation and tried to market a new sedan using a redesigned front ahead of a modified Cord passenger compartment. This proved to be an expensive failure as the Cord components had to be welded together, and the public had lost its patience with Graham. In September 1940, Graham folded, but the facilities were used to produce Weasel amphibious tanks in World War II, powered by the same Continental six. Joseph Frazer bought the automaking assets in 1944 to build the short-lived postwar Frazer automobile. Interestingly, postwar Frazer and Kaiser cars kept that venerable flathead I-6, and Kaiser offered some supercharged models, arguably bringing things full-circle.

The rare sight of a Graham Sharknose today seldom fails to elicit stares and smiles. *Curbside Classics* Editor Paul Niedemeyer said it best when he wrote that Amos Northup "jumped the shark" with the Spirit of Motion.[3]

—KNG

NOTES

1 "Graham by a Nose," by Michael Lamm, *Special-Interest Autos*, October, 1970, pp. 40-43.

2 "Spirit of Motion," by David LaChance, *www.hemmings.com/magazine10/Spirit-of-Motion/2127561.html*

3 "A Bit Too Far Ahead of Its Time," by Paul Niedemeyer, Curbside Classics, October 19, 2015, pp. 1.

Graham photos by Peter Harholdt

1935 BUGATTI TYPE 57 AÉROLITHE

COLLECTION OF CHRIS OHRSTROM

The Bugatti Aérolithe was arguably the most sensational automobile of the mid-1930s. The fact that it was a Bugatti should have been no surprise. Ettore Bugatti himself sprang from a remarkable Italian family of artists. His father, Carlo, created intricate, museum-quality Art Deco furniture. A younger brother, aptly named Rembrandt, was a renowned sculptor of animals. A trained engineer with an artistic soul, Ettore Bugatti and his family lived regally on a lavish estate in Alsace-Lorraine, near their factory in eastern France. Jean Bugatti shared his father's passion and became a highly imaginative automobile designer.

From 1911 to 1939, Bugatti only produced about 12,000 automobiles, ranging from fast sports and touring cars to successful racing models and even regal limousines. Bugattis were technically complex, often temperamental, very expensive, and hauntingly beautiful. Ettore experimented with aerodynamics and the use of exotic, lightweight metals like magnesium. Known as "Le Patron" (the Boss), he favored technology like overhead camshafts, multi-valve engines, and self-adjusting de Ram shock absorbers. But he could also be conservative. Bugatti eschewed supercharging at first and clung to cable-operated brakes long after hydraulics had proved superior.

The Depression of 1929 was slow to impact France because of that country's high tariffs and restricted trade policy, but by the early 1930s, the luxury automobile market had dwindled. Ettore and Jean knew that a special new model was needed to help their company survive. The Type 57, introduced in 1934, was that car. Its styling was contemporary, and custom coachwork was available for those with means.

To maximize the Type 57's impact, Bugatti introduced a streamlined sports model at the back-to-back 1935 Paris and London Motor Shows. Initially called the Compétition Coupé Aérolithe, the avant-garde speedster was beautifully curved from every angle. Its flowing architecture was a marked contrast to square-rigged cars of the era. The Aérolithe (French for meteor) rode on a Type 57 chassis, with gondola-shaped frame-rails that tapered rearward for an aerodynamic appearance. The engine was a normally aspirated 3.3-liter, DOHC straight-8.

Competitive makes experimented with low-volume aerodynamic models, like the sinister-looking Mercedes-Benz 500K/540K Autobahn-Kurier and Talbot-Lago's voluptuous Teardrop coupes, but nothing on European roads in that era was as outrageous. The Aérolithe was an overnight design sensation, but orders didn't pour in.

The show car was fabricated from Elektron, an expensive magnesium and aluminum alloy. That metal proved difficult to weld, so Jean Bugatti, assisted by head draftsman Joseph Walter, united the major sections using rivets. A spine-like center rib divided the svelte body, a theme repeated in its teardrop-shaped fenders. The doors cut into the roof, opened forward, "suicide-style," and stopped at the midpoint of the body. They looked sexy but made access a challenge. Marque experts believe Bugatti built two "Aéro" prototypes, but they did not exist at the same time. After the continental show circuit, the cars were dismantled, and some parts were used to build an updated, more civilized version, called the Atlantic.[1]

Four production Type 57 Atlantics, three of which survive today, were hand-fabricated in aluminum. The rivets were no longer needed, but they looked exotic, so the illusion of a riveted spine was retained. Close-coupled, cramped, poorly ventilated, and rather impractical, the sensuous, lightweight coupe was an enthusiast's delight.[2]

For competition, a Type 57 on an ultra-low 'S' chassis was fitted with streamlined, open coachwork. The factory proudly advertised its successes, which included averaging 135.45 mph for one hour, 123.8 mph for 2,000 miles and 124.6 mph for 4,000 kilometers. A Type 57 won the 24 Hours of Le Mans in 1937. An avid horseman, Le Patron was convinced automobile competition improved the breed, as it did with thoroughbred racing.

Using the lines of the Aérolithe-inspired Atlantic, Jean Bugatti designed the Atalante, a slightly larger, more comfortable, production Grand Tourer on the Type 57 chassis. More successful than the Aéro or the Atlantic, about 40 Atalantes were built on the standard Type 57, and the sporting Type 57 chassis before the War halted all production.

This Aérolithe re-creation was built by David Grainger at The Guild of Auto Restorers, in Bradford, Ontario, Canada. Starting with the earliest surviving Type 57 frame, no. 57104, Grainger and his team obtained sheets of magnesium alloy and then developed a technique to form the body panels and weld them together, using a process that could have been employed in 1935. The crème de menthe finish was taken from a painting of the original Aérolithe, done by a Bugatti engineer named Reister. Every detail was meticulously duplicated, including the double-sided Dunlop tires. The result has been welcomed in Bugatti circles and was featured in *Bugantics*, the Bugatti Owners Club quarterly.[3]

—**KNG**

NOTES

1 Bernhard Simon and Julius Kruta, *The Bugatti Type 57S* (Munster, Germany Verlagshaus Monsenstein und Vannerdar, 2003), 36–38.

2 L. G. Matthews, Jr., *Bugatti Yesterday and Today* (Paris: Editions SPE Bathelemy, 2004), 14–20.

3 David Grainger, "Aerolithe," *Bugantics*, Spring 2013, vol. 76, no. 1, 14–21.

Bugatti photos by Joe Wiecha

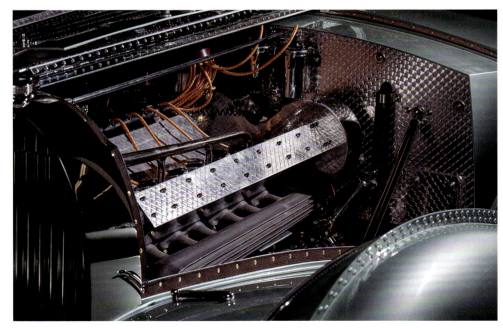

1935 HOFFMAN X-8

COURTESY MYRON AND KIM VERNIS

The true story of this unusual streamlined sedan from the mid-1930s has only been recently discovered. Roscoe C. "Rod" Hoffman was an inventor and engineering contractor in Detroit. Well-known in top automotive circles, Hoffman specialized in chassis and suspension design. A 1911 Purdue University graduate engineer, who'd worked for Haynes and Pierce-Arrow, he started his own company, Hoffman Motor Developments, in 1934. HMC did contract work for General Motors, Studebaker, and Packard.

This unique Hoffman X-8 sedan is a sole surviving prototype that's been a mystery for many years. At first look it is reminiscent of a Chrysler Airflow, but on a slightly smaller scale. And some of its features remind one of a Tatra, particularly its rear-mounted, 8-cylinder engine. But that's where the resemblances end.

The Hoffman's streamlined—if a bit chunky—shape is similar to period experimental cars and even a few production autos from its era. But for years it was unclear who had commissioned it. Researchers long believed that Henry Ford might have been involved, in conjunction with Ford's French Mathis subsidiary. Hoffman was apparently sworn to secrecy, so even when he gave the car to industrial designer Brooks Stevens in 1961, he didn't divulge much information on the X-8's origins. But an old contract has surfaced from the Norwalk Company, in Norwalk, Ohio. It seems that the Fisher brothers established Norwalk as a shell company that authorized and financed Hoffman to build two prototypes.

It's believed that the Fishers, whose giant firm built bodies for General Motors, had considered becoming full-fledged auto manufacturers. They even planned to buy Hudson to realize that plan. According to the Hoffman's present owner,

Myron Vernis, the contract with Rod Hoffman was initially for $65,000, and it grew to $168,000 as a two-car project ensued, consisting of the X-8 and a smaller car.

Secrecy was key, not just to keep the unusual cars themselves hidden until launch, but because the Fishers didn't want speculators to invest in Hudson stock and drive up the price. But the word got out. When a European investor began buying up Hudson shares, the Fisher brothers abandoned their quest to acquire Hudson. It's not known if a second prototype was ever built.[1]

The suspected connection with Henry Ford—which isn't true—derived from his patents for rear-engine cars from 1934 to 1939, and his experiments with X-8 engines. Ford would have built *anything* but a six, because his hated rival Chevrolet offered them. On the way to the 1932 Ford V-8, Henry experimented with several X-8 configurations. But his engines were flatheads (side-valves); the 168.4-cid X-8 in the Hoffman is a highly advanced overhead cam setup with unique combustion chambers, reminiscent of the asymmetrical "flame path" designs popular with drag racers in the 1960s.

The X-8 engine's intake ports are cast into the block. The only production parts are a pair of Ford V-8 water pumps and a conventional 2-bbl. Stromberg carburetor. The crankshaft, according to the present owner, "has three main journals, with the center journal locating four rods that are opposed 180 degrees. It should shake itself apart," Vernis says, "but it's totally smooth without the use of any counterbalance."

So who built the Hoffman and where? Vernis has rare photographs of the Hoffman's unitized body being constructed by the Budd Company (a major supplier to the US auto industry) in Detroit. In one of the photos, well-known engineer Joseph Ledwinka appears. Budd and

Ledwinka pioneered unit construction, and they worked together on Citroën front-wheel drive cars. Interestingly, Joseph Ledwinka's cousin, Hans Ledwinka, designed the Tatra T77a in this exhibition.

The Hoffman X-8 rides on a 115-inch wheelbase and weighs about 3,100 lbs. It has an all-steel, unitized body and frame, with honeycomb floor-perimeter strengthening members. The car's front suspension consists of a tubular axle, transverse leaf springs, front trailing arms, and tubular shocks. The rear suspension, quite advanced for an American car of that era, features fully independent half-shafts with Cardan universal joints at each end, along with longitudinal leaf springs and trailing arms. The brakes are four-wheel hydraulic drums.

Many rear-engine cars were especially designed to capitalize on the advantages of streamlining. John Tjaarda, working for the Briggs Body Company, who designed the Sterkenburg rear-engine sedan, and William Bushnell Stout, who developed the Stout Scarab, were keen proponents of rear-engine "pusher" designs in the mid-1930s. Hoffman was in good company with his ideas. Mike Lamm quoted former Packard stylist John Reinhart, who said, "Rod (Hoffman) used to be very famous for designing rear-engine cars. In fact, we used to call him, 'Rear-Engine Hoffman.'"[2]

The Hoffman is the only car ever built around an X-8 engine configuration—and it's simply one of a kind.

—KNG

NOTES

1 "Supreme Whatsit," by Mike Lamm, *Special-Interest Autos*, Sept-Oct., 1974. pp. 52.

2 ibid., pp. 54.

Hoffman photos by Peter Harholdt

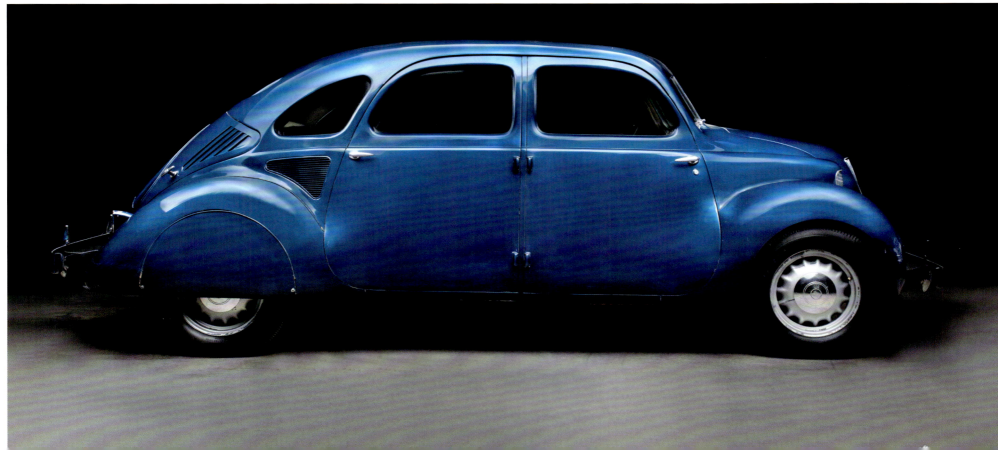

1936 CORD 812SC WESTCHESTER SEDAN

COURTESY KEVIN CORNISH

A sensation when it went on sale in 1936, the Cord 810 is one of the most beautiful cars ever built. Designed by Gordon Miller Buehrig, who penned many memorable models for Auburn and Duesenberg, the car was named for Cord Corporation owner and famed entrepreneur Errett Lobban Cord. Originally intended to be a "baby" Duesenberg, its many advanced features included front-wheel drive, fully independent front suspension, a unitized body with pontoon fenders and an "alligator" hood, a 125-bhp, 90-degree, 288-cid V-8 engine with aluminum heads (built by Lycoming, an aircraft company owned by Cord), and a four-speed, pre-selector Bendix transmission with vacuum/electric shifting. The horn ring, covered gas filler cap, and crank-out, "hidden" headlights were all American firsts.

The Cord's sleek, streamlined styling, arguably more than any other car of its era save the Auburn Model 851 Speedster from the same company, celebrated the Art Deco style. The 810's unique coffin-shaped hood incorporated horizontal rows of slatted cooling louvers on three sides. Its curvaceous fenders and exquisitely arched roofline shouldn't have worked so well with the angular hood, tall waistline, and low windows, but the overall effect was very pleasing to the eye—a marked contrast to boxy rival models. Buehrig used the same stamping die for all four doors to save tooling costs, and the resulting symmetry resulted in a perfectly-shaped cabin. Practical considerations (i.e., buyers demanding more storage space) later forced the factory to design a fussy bustle-back luggage compartment.

Rushed to market in the midst of the Depression, the stunning Cord's development work was not quite finished when it went on sale. Show cars had been displayed without transmissions, pending final development. Orders poured in, but early examples suffered from overheating, balky gearboxes, and braking problems. Priced around $3,000, a new Cord cost more than some Cadillacs, but it was perceived as a smaller car. In 1937, when this lovely Model 812 sedan was produced, prices were raised nearly 20 percent. A Schwitzer-Cummins centrifugal supercharger, good for an advertised 170 bhp, was optional. Tests showed the output was closer to 195 bhp. Supercharged models with chromed exhaust pipes emerging from the hood mimicked blown Auburns and Duesenbergs, adding to the Cord's cachet.

In addition to engineering advances and aerodynamic styling that were years ahead of its contemporaries, the Cord was fast for its era. Testers for *The Autocar*, a respected British magazine, reached 102.27 mph on a top-speed run. Seeking favorable publicity at home, Cord targeted the Stevens Challenge Trophy, awarded to the fastest closed production car over a 24-hour time period. Samuel Stevens, the namesake of the competition, who'd competed in the early Vanderbilt Cup Races, was convinced that racing "improved the breed," and believed that a car's durability was best measured when it could sustain its top speed over an extended time period.

By 1937, although Auburn and Duesenberg were finished, struggling Cord's management felt that a record-setting performance might lend momentum to the company's flagging sales. Stevens provided a handsome silver trophy to be presented to the American automaker whose closed-bodied stock car, certified by the AAA and running on pump gasoline, achieved the highest average speed at the Indianapolis Motor Speedway over a timed 24-hour period. Famed endurance racer David Abbott "Ab" Jenkins employed two supercharged Cord 812s in a record attempt designed to focus attention on Cord's performance and durability. Although one Cord suffered a suspension failure, Jenkins completed the 24 hours in the second 812, averaging 79.577 mph.

With the damaged Cord repaired, both Model 812s were taken to the Bonneville Salt Flats in July 1937 to compete for flying start records in Class C and the Unlimited Class. Jenkins completed nearly 2,500 miles in 24 hours, averaging more than 101 mph, including stops for fuel and tires. Sadly, the supercharged Cord's record performance didn't revive the company's failing fortunes. Some 2,930 Model 810s and 812s, including convertible coupes, phaetons, and long-wheelbase Custom Berlines, were sold in just two years of production. While the Cord Model 810/812 is recalled today for its engineering prowess, its stunning Art Deco styling earned it a place in the 1951 Museum of Modern Art "8 Automobiles" exhibition, where curator Arthur Drexler called it "a solemn expression of streamlining."

—KNG

In addition to engineering advances and aerodynamic styling that were years ahead of its contemporaries, the Cord was fast for its era.

Cord photos by Peter Harholdt

1936 STOUT SCARAB

COLLECTION OF RON SCHNEIDER

William Bushnell Stout's accomplishments included founding *Aerial Age*, the country's first aviation magazine, in 1912. An accomplished technical journalist and engineer, Stout designed the lightweight Imp cyclecar for the Scripps-Booth Company, which led to a sales management job at Packard, and then the chief engineer's position in Packard's fledgling aircraft division. Stout's advanced work on internally braced, cantilever-wing aircraft pioneered the development of all-metal airplanes and resulted in the Stout Air Sedan. After Henry Ford purchased Stout Engineering in 1924, Stout's 3-AT aircraft design evolved into the sturdy Ford Tri-Motor. In 1929, Stout Air Services, a Midwest-based passenger airline, was purchased by United Air Lines.

Stout began creating a radical sedan concept in the early 1930s. Drawing on his extensive aeronautical background, Stout believed the use of aircraft construction techniques would result in a futuristic car that would go faster and be more economical than a conventional auto. He envisioned a smooth, slightly tapered, and, for its era, startling shape. A tubular frame covered with aluminum panels surrounded the Scarab's rear-mounted, flathead V-8 engine. A Ford three-speed transmission with a custom transfer case and a six-row chain powered the rear wheels. In what has since become modern practice, all four wheels were located at the corners of the vehicle. Stout's front independent coil spring suspension design anticipated today's MacPherson strut setup. The rear suspension was composed of twin transverse leaf springs.

The Scarab's roomy passenger compartment was positioned within the car's wheelbase. Access to the interior was through a central door on the right side, and there was a narrow front door on the left for the driver. The driver's seat and a wide rear bench seat were fixed. Other seats could be repositioned so the front could accommodate three passengers across, or moved so passengers could sit around a small table. Stout's unusual configuration anticipated the first minivan. The interior was trimmed in wood, the headliner was varnished wicker, and, as in the Chrysler Airflow, the seats were accentuated with chrome rails, with the seating surfaces finished in leather.

The Scarab's distinctive turtle-shell styling celebrated its Art Deco influence, beginning with decorative "mustaches" below the split windshield, including the unusual-shaped

Stout's unusual configuration anticipated the first minivan.

headlamps covered with thin grilles, and culminating in fan-shaped vertical fluting that framed elegant cooling grilles. Its Ford V-8 engine was reverse-mounted, with the cooling fan in the rear. The Scarab's smooth envelope body enclosed the wheels and contributed to quiet operation. The name was a not-too-subtle underscoring of the sedan's pleasant, beetle-like shape. Several period tin toy manufacturers copied the Scarab's shape.

Even more than the ill-fated Chrysler Airflow, the Stout Scarab was radically different from contemporary vehicles. At $5,000, it was very expensive as well. The Depression-wracked buying public did not recognize its many advantages when a Ford DeLuxe "Fordor" could be had for $650. A few designers developed similar configurations, most notably John Tjaarda, with his Sterkenburg experimental car; Czechoslovakia's Hans Ledwinka, with the air-cooled, rear-engine Tatra; and Ferdinand Porsche, whose design study for a "people's car" became Germany's Volkswagen. With modifications, Tjaarda's concept became the streamlined Lincoln-Zephyr; Ledwinka's and Porsche's designs also saw volume production. But Stout's venture failed.

Stout boldly advertised in *Fortune* that production would be limited to 100 cars. A few of his investors and board members, like William Wrigley, the chewing gum magnate, and Willard Dow, of the Dow Chemical Company, purchased Scarabs. Other owners included tire company owner Harvey Firestone, Robert Stranahan of the Champion Spark Plug Company, and radio host Major Edward Bowes. It's believed that at least six Scarabs were built; some sources say nine. The cost to produce them was greater than the already-high $5,000 price tag. Five are known to survive.

Ron Schneider, the lender of this car, owns two Scarab sedans. He says both of his cars were driven extensively by earlier owners and states this car was driven more than 150,000 miles before he acquired it. He entered the 1989 Great American race and repeated the feat the following year. Schneider believes this is the Robert Stranahan car.

A forerunner of the modern minivan, with its aerodynamic shape and reconfigurable interior, William Stout's clever Scarab sedan was well ahead of its time and remarkably like its modern counterparts.

—KNG

Stout photos by Peter Harholdt

1937 AIROMOBILE

COURTESY OF THE NATIONAL AUTOMOBILE MUSEUM, THE HARRAH COLLECTION

For decades independent automakers have been fascinated with the notion of a flying car. A few companies have made the concept work, with functioning prototypes and even short production runs, but the buying public has never really been interested. That said, the airplane—which was considered a highly advanced invention by the 1930s—influenced many automakers with its inherent streamlining and lightweight construction. Minimalist good looks and distinctive shapes—even wings and fins—all lend themselves well to the concept of a futuristic car, even if actual flight is not among a vehicle's goals. William Bushnell Stout's Scarab is a good example. Stout, an acclaimed aircraft designer, with the Ford Tri-Motor among his many credits, used a number of aircraft principles to build his futuristic Scarab sedan, although it was never intended to resemble an airplane.

The Airomobile was the brainchild of American Paul M. Lewis. He chose a somewhat unorthodox tripod layout for the inherent torsional strength it would lend to his car's chassis. The three-wheel layout enjoyed some popularity in England at the time; owners of the Morgan three-wheeler and rival vehicles saved on licensing fees by utilizing a tax law that classified three-wheeled vehicles as motorcycles. That was not Lewis's goal with his design, however. Rather, Lewis planned an affordable, lightweight, aerodynamic, and distinctive-looking car of the future, circa 1934.

To bring his dream car to life, Lewis enlisted some serious help. The first Airomobile renderings were done by John Tjaarda, a brilliant designer at Briggs Body Company, whose streamlined, four-door, rear-engine Sterkenburg concept car was later adapted by E.T. "Bob" Gregorie to become the 1936 Lincoln-Zephyr. Carl Doman and Ed Marks, former Franklin Automobile Company engineers, had established

a freelance engineering and prototype production company, Doman-Marks. They translated Tjaarda's renderings into a workable engineering plan for a production three-wheeler.

The result was an aircraft-shaped car with a 180-inch-long "fuselage." Its sheet-steel body panels were fastened to a lightweight steel frame made of round tubing; the chassis consisted of a perimeter frame made of square tubes. The engine rested on a removable front subframe. The Airomobile's independent front suspension was composed of

via Spicer constant velocity joints that were derived from proven Citroën front-wheel drive components. Curiously, the car's single rear wheel was not equipped with a brake, which must have made for scary panic braking.

Only one prototype Airomobile car was ever constructed by Paul Lewis's company, Lewis-American Airways, and this is it. In 1937, the Airomobile (it was originally called the Airmobile) was projected to sell for $550. By way of comparison, a 1937 Ford Tudor sedan was $579. Lewis

Minimalist good looks and distinctive shapes—even wings and fins—all lend themselves well to the concept of a futuristic car, even if actual flight is not among a vehicle's goals.

tubular shock absorbers, coil springs and control arms. The odd car's single rear wheel, which was smaller than the two front wheels, was supported by a longitudinal, semi-elliptic leaf spring, a lone trailing arm, and a single hydraulic shock absorber. A relatively long, 126-inch wheelbase reportedly provided acceptable ride and handling characteristics. The Airomobile weighed 2,200 lbs.

As if this car's tri-cornered stance weren't unusual enough, Lewis planned to power the Airomobile with a Skinner Motor Corporation sleeve-valve, horizontally opposed four-cylinder engine. When that engine was not ready for production, he substituted a then-new, air-cooled overhead-valve flat four. It displaced 129-cid and developed 57 bhp at 3,700 rpm. The same unit later powered the White Motor Company's "White Horse" delivery vehicle. Other similar engines, built with aluminum cylinder blocks, were used as aircraft power plants. The flat four, which resembled engines used by VW and Porsche, powered the front wheels

boldly drove his three-wheeled car all over the United States to demonstrate it, completing some 45,000 miles in less than a year and averaging an impressive 43.6 mpg.

To entice buyers, the prototype was restyled and the front suspension was re-engineered in 1938, but to no avail. In spite of Lewis's efforts, he was unable to attract sufficient financial backing to put his curious streamlined car into production. The prototype Airomobile was later acquired by the now-defunct Harrah's Automobile Collection.

Alongside competitively priced cars of its day, the Airomobile certainly attracts attention. The diminutive car's lack of wings and vertical tail (they really weren't needed), and its unusual, aircraft-inspired shape, mark it as an oddity, despite its streamlined influence.

—KNG

Airomobile photos by Jeff Dow

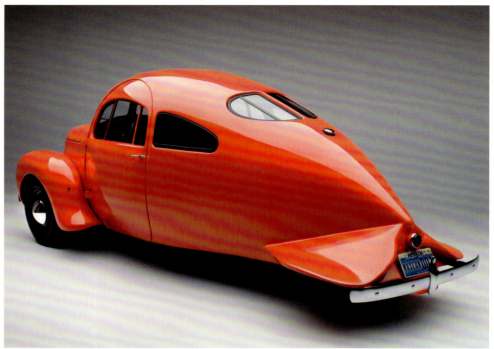

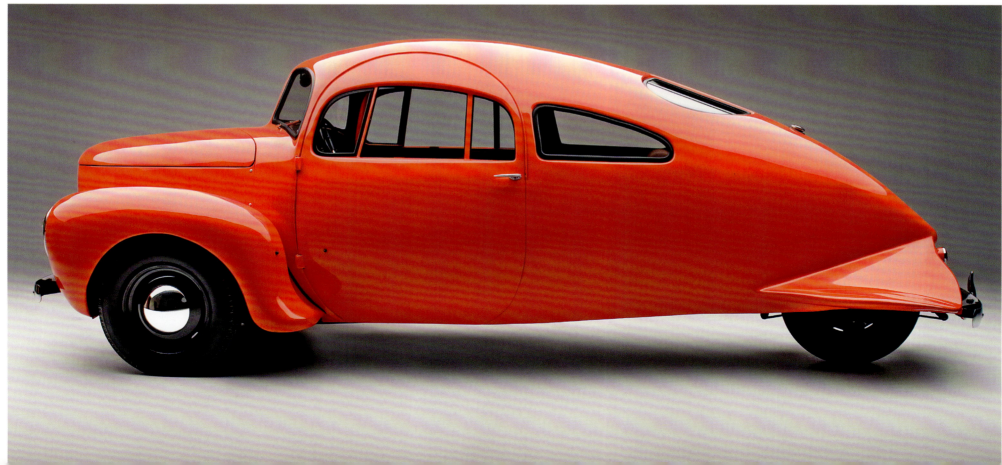

1937 LINCOLN-ZEPHYR COUPE

COURTESY ALAN JOHNSON

Arguably one of the loveliest American cars of the mid-1930s, the Lincoln-Zephyr was more than a modest sales success. Enthusiasm for its flowing lines conclusively proved that Americans (and others) would buy streamlined automobiles in significant numbers.

It took courage for Edsel B. Ford and E.T. "Bob" Gregorie to offer their clientele a streamlined car in the Lincoln lineup. Previous classic Lincolns, while elegant, expensive, and often coachbuilt, were somewhat staid designs. With its sales drying up, Lincoln badly needed a boost with a new model that would capture the wealthy public's imagination. Luckily, the year 1936 saw the United States just emerging from the Great Depression, so the timing for a new approach was right.

The Lincoln-Zephyr design emerged from an unlikely source. John Tjaarda, a designer at Briggs Manufacturing Company, had produced a radical, rear-engine prototype car he called the Sterkenburg, named for his Dutch family's ancestral home. With its low silhouette, tapered, teardrop-shaped fuselage, skirted rear wings, and a prominent tail fin for stability, it closely resembled Hans Ledwinka's 1934 Tatra T77. But the Sterkenburg had appeared a year earlier in 1933 when, at the request of Edsel Ford (a Briggs client), Tjaarda showed several advanced designs that could be either front or rear-engine cars.[1]

Mr. Ford, who possessed wonderful artistic sensibilities, realized that the Sterkenburg had the potential to be the next Lincoln, albeit in front-engine form. He asked Tjaarda to redesign the prototype's rounded front end. An avowed proponent of rear-engine designs, Tjaarda categorically refused. Gregorie, however, then just 24 years old but the de facto head of the group that would become Ford, Mercury,

and Lincoln's styling and design department, understood how to give Edsel what he wanted. He produced a one-piece, rear-hinged "alligator" hood and a tall, handsome vee'd grille that perfectly complemented (and retained) the bold, aerodynamic shape of Tjaarda's experimental sedan.[2]

The new car's name came from a crack streamlined passenger train of that era: the *Burlington Zephyr*. At the 1934 Chicago World's Fair, the sleek Zephyr helped to popularize public interest in *Streamline Moderne* design. Soon streamlined products began to come to market, items like radios, vacuum cleaners, mixers, and even pencil sharpeners. Such devices hardly needed to reduce air resistance to function properly, but they now looked very modern.

Tjaarda's original concept for the Sterkenburg incorporated a strong, fully unitized body. That idea carried over to the production Lincoln-Zephyr, making it the first production car so equipped in North America. Ford engineers extended the venerable L-head V-8 into a 268-cid V-12. At just $1,320 for the four-door sedan, it was the lowest-priced twelve in the industry, and it sold for less than half the price of a traditional Lincoln. The three-passenger coupe, a masterpiece of curvaceous styling, first appeared in 1937, and Zephyr sales doubled that year. For 1938, Gregorie completely redesigned the tall, narrow grille to be split in two parts, positioned low and horizontally, to cure the engine's tendency toward overheating. That doubtless vexed GM design boss Harley Earl (who immediately recognized its perfect blend of function and beauty), and it simply revolutionized the look of future American cars.

By 1939, the Zephyr grille evolved even further. Gregorie, who had trained as a yacht designer, now incorporated subtle nautical elements into the Lincoln look. Edsel Ford loved

the Zephyr and asked Gregorie to design a smart-looking convertible on the Zephyr chassis that he could use to impress fellow wealthy snowbirds in Palm Beach, where he and his family wintered each year. The result was the Lincoln Continental, a design triumph that encouraged a flow of orders, helping to ensure Lincoln would weather the war and emerge strongly in the late '40s. Historian Beverly Rae Kimes noted, "the outstanding nature of the Lincoln-Zephyr's styling was further demonstrated by its transformation into one of America's most beautiful Classics, the Continental."[3]

Although John Tjaarda might never have imagined it, the Lincoln-Zephyr, spawned by his futuristic, experimental Sterkenburg sedan, eventually sold more than 180,000 units, finally ending production in 1942. Arthur Drexler, Curator of Architecture at The Museum of Modern Art in New York City, and curator of the critically acclaimed 1951 MoMA exhibition, "8 Automobiles," called the Lincoln-Zephyr, "the first successful streamline car in America."[4]

—KNG

NOTES

1   *The Old Car Book*, John Bentley, Fawcett Publications, Greenwich, CT, 1953, pp. 136.

2   *Curves of Steel*, Phoenix Art Museum Catalog, Jonathan A. Stein, Editor, Coachbuilt Press, Philadelphia, PA, 2007, pp. 50-53.

3   *Standard Catalog of American Cars*, Beverly Rae Kimes and Henry Austin Clark, Iola, WI, 1989, pp. 846

4   *8 Automobiles*, The Museum of Modern Art Catalog, New York, NY. 1951. pp. 4.

Lincoln-Zephyr photos by Dale Moreau

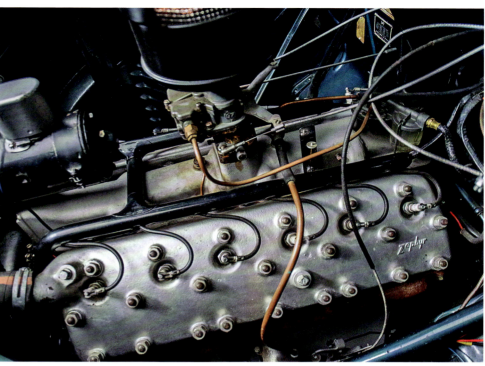

1938 DELAHAYE 135M ROADSTER

COLLECTION OF THE PETERSEN AUTOMOTIVE MUSEUM, GIFT OF MARGIE AND ROBERT E. PETERSEN FOUNDATION

Joseph Figoni and Ovidio Falaschi, renowned Paris-based coachbuilders throughout the classic era, were noted for their swoopy, elegant, custom coachwork in the mid- to late 1930s, especially on Delahaye and Delage chassis. Often described as "Paris gowns on wheels," or "rolling sculpture," Figoni & Falaschi's luscious creations won numerous concours d'elegance. The carrosserie's exclusive clientele included royal notables like Emperor Bao Dai of Annam, the Maharanee of Kapurthala, and the Marchioness of Cholmondeley.

At the 1936 Paris Salon, Figoni & Falaschi presented a marvelous Delahaye Type 135 roadster on the Competition Court chassis, with fully enclosed front fenders, impossibly low, integrally mounted headlamps, a rakish split windscreen, and a 3-carburetor, 3.5-liter six-cylinder engine. The design was a close derivative of a fanciful sketch by noted French motoring artist Geo Ham (born Georges Hamel). The price was 150,000 francs (about $27,000), and the lovely car was purchased right off the show stand by Prince Aly Khan, a wealthy and notorious playboy.

French coachwork authority Richard Adatto wrote, "Joseph Figoni took modern streamlining to the next level by allowing the optimal aerodynamic shape to dictate the styling. Instead of pontoon fenders that protruded from the car's body, Figoni found a way to incorporate them into the body, heightening the impression of a singular, flowing form."[1]

It is believed that just 11 of these voluptuously bodied Salon de Paris roadsters were built, on short- and long-wheelbase Delahaye chassis. This 135MS, chassis number 49169, has a 3.5-liter, 6-cylinder engine. There are several subtle styling differences among the 11 roadsters, including slanted hood louvers on some, scalloped air vents on others, and varying complements of painted and plated trim.

They all have four fully skirted fenders, an audacious but beautiful styling conceit that most critics believed would be impossible to execute. The front fender edges are gracefully bowed outward so that the wheels can turn side-to-side. Figoni & Falaschi found a way to do this and still retain the car's elegance and joie de vivre. Besides Delahayes, a few Talbot-Darracq cars, including a gorgeous roadster, received this curvaceous front-fender approach.

Some Salon de Paris roadsters, like this one, had split windscreens. Others featured windshields that could be folded flat, and still others retracted into the cowl for a low and racy appearance. Most of these cars were ordered with advanced Cotal electromagnetic preselector transmissions. Lacking fully automatic gearboxes, the preselector Cotals—and similar units made by Wilson—allowed the driver to preselect a gear with a tiny lever and then shift smoothly once the clutch was depressed. They were complex and expensive, so not all Delahayes had them. This roadster is equipped with a conventional four-speed manual gearbox. The front suspension is independent with upper wishbones and a lower transverse leaf spring; the rear is a live (solid) axle.

A long-wheelbase model, this elegant two-seater was ordered by a Monsieur Fould of Oran and delivered to him in Algeria. Miraculously, it survived World War II, but then it disappeared in the 1950s. Discovered in a remote farmyard in the Algerian mountains in 1992, it was virtually complete, with its original engine, Carrossier Figoni et Falaschi shields and chassis plate, and lacking only the aluminum decklid. The long-lost Delahaye's lucky buyer paid the ridiculously low sum of just 60 British pounds for it! It is possibly the all-time "barn find" bargain.

Completely restored over several years by British experts Crail of Southall, Middlesex, and Carrosserie Fernandez of Switzerland, the Delahaye is largely period-correct except for the substitution of postwar Solex carburetors for the sand-encrusted prewar units that were found with the car. It was subsequently sold to Robert E. Petersen at the Brooks Auction in Monterey, California, in August, 1999, for $1.25 million. A similar Geo Ham Delahaye 135M roadster, on the short chassis and owned by the late Jacques Harguindeguy, won Best of Show at the Pebble Beach Concours d'Elegance in August 2000.

—KNG

[Its] four fully skirted fenders [were] an audacious but beautiful styling conceit that most critics believed would be impossible to execute.

NOTES

1 Ken Gross and Peter Harholdt, *Sensuous Steel: Art Deco Automobiles* (St. Paul, MN: Stance & Speed), 78.

Delahaye photos by Scott Williamson

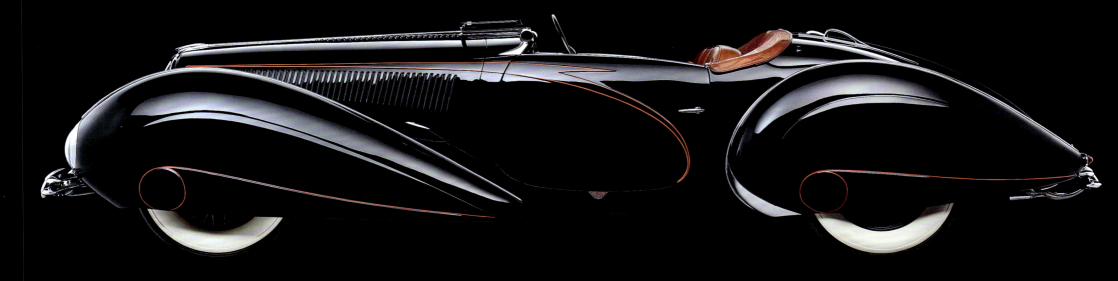

1938 MERCEDES-BENZ 540K STREAMLINER

MERCEDES-BENZ MUSEUM

For years, faded black-and-white photos of a streamlined 540K Mercedes-Benz coupe tantalized historians. Seemingly lost for all time, little was known about this unique car, save that it was specially built as a test vehicle for the German branch of the Dunlop tire company. Its designer was Hermann Ahrens, who penned the elegant 540K Special-Roadsters built by Sindelfingen, the custom body branch of Daimler-Benz AG in Stuttgart.

Mercedes-Benz archivist Gerd Langer unlocked the puzzle in 2006 and 2007, when he found a reference, entry no. 284, in a list of some 900 vehicles in the vast M-B storage halls. Langer's search turned up an unrestored W29 chassis equipped with a rear swing axle with finned brake drums, one road wheel, and a few other components. The chassis plate, still attached, bore the number 189399. Luckily, the company's commission books, voluminous hand-written records of every Mercedes-Benz built in that era, had been preserved. A listing indicated that this car was registered as IT1346901 and sold to Dunlop Germany on June 14, 1938, after which it was used for tire testing and subsequently, in 1940, converted to run on liquid petroleum gas.

Speaking with *Thoroughbred & Classic Cars* editor Phil Bell, Langer recalled, "I found the original design drawing, pen on parchment, the Linienrisszeichnung—a microfiche printout of the original drawings—details of the silver paint and grey leather interior, and six original photographs of the car in a scrapbook of special design projects belonging to in-house stylist Hermann Ahrens." With this, Langer began to solve the mystery.

Originally, the streamlined coupe (or Stromlinienwagen, in German) was reportedly developed for a proposed Berlin-Rome road race, intended to rival Italy's Mille Miglia. The race was canceled in 1939, due to hostilities. Dunlop Germany acquired the streamliner and used it to test tires at high speeds on the autobahns. Langer found the number 2.90:1 stamped on the rear axle casing. With the 540K's supercharged straight-8 pegged at 3,600 rpm, such tall gearing would permit a top speed of 185 kph or about 115 mph (the 540K's standard axle ratio was 3.08:1). Little more is known about the streamliner, except that it was used by a U.S. serviceman after the war, then returned to Mercedes-Benz in the 1950s. After this date its bodywork was removed and most of its mechanical components were taken off and lost.

Following the chassis's rediscovery, restoration technicians at Mercedes-Benz carefully cleaned and repaired the chassis, then covered it with a thin layer of wax to preserve its time-honored patina. They elected to build a new body using traditional methods. Ralph Hettich, who led the restoration team, told Phil Bell that, "...the hardest part of recreating the body was using the Linienrisszeichnung, a very complex technical drawing. A craftsman of the thirties would have no problem reading it," he said, "but that knowledge is not so common now." The team studied the document and converted it to a modern CAD (computer-assisted design) data set to reproduce the original shape. They had found a piece of the old body shell, made of lightweight duraluminium, so they knew its composition.

As factory records were incomplete, the team was challenged to build the replica body and its ash-framed superstructure. There were few interior photographs. Drawings indicated two folding seats in the rear. Some clues were found, like photos of small ribs used where the bodywork tucked under the chassis to form a belly pan, and small fillets located in the frame for the attachment screws. The split windshields were made in metal at first, in order to replicate their exact shape, before forming them out of glass. Some of the aerodynamic details that delighted the restorers included recessed door handles, a tapered tail, extensive hood louvers to allow trapped air to escape, and even rounded sidelights on the fenders.

The team re-used all the original parts that were found, and meticulously replicated missing pieces. As this streamlined coupe was based on a 540K, its mechanical components were comparatively easier to source. Mercedes-Benz had an original M24 5,401-cc, 540K engine in its collection. Restored and dynamometer-tested, it developed 115 bhp at 3,800 rpm. With the Roots-type supercharger engaged, it developed 180 bhp for short periods.

Photographs existed of the original curved nutwood dashboard, but the cabin's interior fabrication involved calculated guesswork using details they knew from Mercedes-Benzes of that period. The finished car was tested in the M-B wind tunnel in Stuttgart-Untertürkheam, the oldest functioning wind tunnel in Germany. Using a standard 540K coupe for comparison, it was found that the Stromlinienwagen's Cd (Coefficient of Drag) is a remarkable 0.36, a figure not generally achieved until the 1960s. The stock 540K coupe had a Cd of 0.57, a remarkable difference for the era, particularly as the original Stromlinienwagen retained the standard 540K's tall radiator.

The restoration work took two and one-half years. Michael Bock, head of Mercedes-Benz Classic, said, "the car stands today as it would have been seen in public in 1938: a herald of the future, whose high driving speeds and exceptional performance are quite apparent even when it is stationary. A unique, avant-garde messenger from an era still characterized by traditional bodywork styling, which has sped its way straight into the modern age, and still somehow looks up-to-date, even today."

—KNG

Photos courtesy Daimler AG

1938 TALBOT-LAGO T-150-C-SS

COLLECTION OF MULLIN AUTOMOTIVE MUSEUM FOUNDATION

Figoni & Falaschi's Teardrop remains one of the most iconic automobiles to emerge from the age of streamlining. Its sensuous curves and dramatic styling have withstood the test of time, appealing to modern sensibilities just as they appealed to audiences of the 1930s. Ultimately, the Teardrop was a product not of a wind tunnel, but of Figoni's heart, passion, and intuition.

The sporting Talbot-Lago T-150-C chassis inspired open roadsters and closed cars, most notably a series of curvaceous custom coupes. Sensational in their heyday, they remain highly valued today. Streamlined, sleek, and light enough to race competitively, they were called *Goutte d'Eau* (drop of water) and in English they quickly became known as the Teardrop Talbots.

Famed Parisian carrossiers Joseph Figoni and Ovidio Falaschi built twelve "New York-style" Talbot-Lago coupes between 1937 and 1939, so-called because the first was introduced at the 1937 New York Auto Show at the Grand Central Palace, where it was very well received. Imported cars were still a rarity at U.S. shows, and there was nothing from any American manufacturer that remotely matched the Teardrop's curvaceous lines and stunning presence. Five more cars, built in a notchback Teardrop style, were named "Jeancart," after a wealthy French patron. It took Figoni & Falaschi craftsmen some 2,100 hours to complete a body. Figoni & Falaschi patented the Teardrop's distinctive aerodynamic shape, yet no two of these coupes were exactly alike.

To the delight of sportsmen and select coachbuilders, Talbot's president, Anthony Lago, offered a top-of-the-line SS (Super Sport) version on the T-150-C's sturdy ladder style frame, with independent wishbones in front and a live rear axle. The competition engine, designed by Walter Becchia, was a 4-liter six, topped with an overhead-valve head with hemispherically-shaped combustion chambers. Fitted with three Zenith Stromberg EX 32 carburetors, it produced 170 bhp. Some cars were equipped with an innovative Wilson pre-selector gearbox, with a fingertip-actuated lever that permitted instant shifts without the driver having to take his hand off the steering wheel. These lovely coupes were capable of 115 mph. In 1938, a race-spec T-150-C-SS Coupe driven by Jean Prenant and André Morel finished third at the 24 Hours of Le Mans.

This example, the first produced and one of three built with aluminum alloy coachwork, was used in 1937 to publicize the reliability and the elegance of its Figoni & Falaschi coachwork. It is unique in several ways, thanks to its lightweight body, fold-out front windscreen, and competition-style exhaust headers. The single blade, aviation-style bumpers perfectly complement this car's curvaceous shape.

It's unsurprising that such a car has had a notable past. At one point the car crossed paths with Freddie McEvoy, an Australian member of the 1936 British Olympic bobsled team. A prominent player on the Hollywood scene, the dashing Mr. McEvoy's ready access to A-list celebrities like Errol Flynn made him an ideal concessionaire for luxurious automobiles. Reportedly, McEvoy was dining with heiress Barbara Hutton when he bragged that he could drive this very car from Paris to Cannes in less than 10 hours. In spite of the treacherous roads of the Alps and numerous small villages in between, he covered the 565 miles in this Talbot-Lago in 9 hours and 45 minutes, keeping his word and winning a $10,000 bet. At that time, before commercial air travel, wealthy people were always on the lookout for a better way to travel from Paris to the grand hotels on the Riviera. Freddie McEvoy's fast trip caught the attention of this elite group and did much to further the reputation of Talbot-Lago, and Figoni & Falaschi.

This car is chassis 90106. It was originally owned by Woolf Barnato, a famed "Bentley Boy," and later the chairman of Bentley Motors. He saw the car when it was presented at the 1937 London Motor Show and is thought to have bought it straight off the stand.

It's not certain what happened to the car after Barnato's ownership. After spending two decades in England, it was sold in the 1960s to importer Otto Zipper, who brought the car to the U.S., where it was then displayed at the Briggs S. Cunningham Museum in Costa Mesa, California. It was subsequently sold to John Calley, who owned it for two years, and then to Pat Hart, who undertook a comprehensive, multi-year restoration. The Teardrop was purchased by Peter Mullin in 1985.

Shown throughout the world, this Talbot-Lago T-150-C-SS has been recognized with class awards at the Pebble Beach Concours d'Elegance (Moët et Chandon cup for the Best French car) in 1984 when owned by Hart. Mullin improved the car even more and continued to collect Pebble Beach honors, again in 1990, 2000, 2005, and 2008. It won Best of Show and Most Elegant Car at the 2001 European Concours at Heidelberg. It has received technical innovation, styling, and design awards from the Art Center College of Design, the Marin Sonoma Concours, the Rodeo Drive Concours, and it has garnered people's choice awards on several occasions.

As the epitome of aerodynamics, speed, design, and elegance, the Teardrop Talbot-Lago does not have a single bad angle. Noted designer Strother MacMinn once said of the car, "... the Talbot Goutte d'Eau coupe represents what may be one of the finest examples of assembled form ever applied to the automobile."

—RSA/KNG

Talbot-Lago photos by Michael Furman

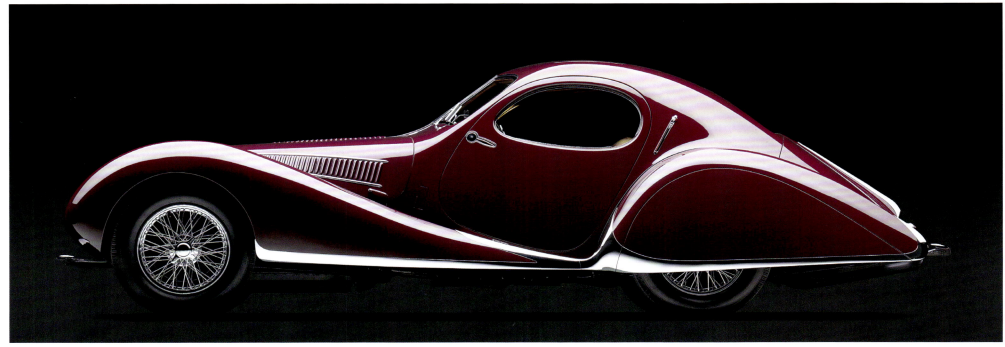

1938 TATRA T77A

COLLECTION OF HELENA MITCHELL AND JOHN LONG

One of the most advanced automotive designs of the mid- to late 1930s came from Czechoslovakia. Founded in 1919 as Czech-based Koprivnicka vozovka, the company evolved into the Nessendorfer Waggonfabrick, and was renamed Tatra in 1927, after the country's prominent mountain range. Tatra vehicles were respected for their innovative engineering and high quality. The engineer largely responsible for that reputation was Hans Ledwinka, who had earlier worked under automotive and aircraft pioneer Edmund Rumpler. Considered "one of the most original and logical thinkers ever to work in the automotive industry," Ledwinka was an early proponent of air-cooled, rear-mounted engines, rigid backbone chassis, and independent suspension, which was advantageous on the rough roads of newly formed Czechoslovakia.[1]

The car in which Ledwinka first deployed these advanced engineering features was the Model T11, and it proved extremely rugged and reliable. In 1926, after 3,540 examples had been sold, the T12 followed and pioneered four-wheel brakes. Production of the T12 was more than double that of its predecessor.

In 1930, Ledwinka and Tatra design engineer Ehrich Ubelacher started to experiment with a rear-mounted, air-cooled, twin-cylinder engine configurations, again using a swing-axle-equipped backbone frame.

The subsequent V570 never saw production, but it was a useful exercise in defining a new Tatra design philosophy. This comprised a rear-mounted engine, a rugged, yet lightweight and simple chassis layout utilizing a pressed steel platform with a central tube for added strength, and a spacious and comfortable passenger compartment located within the car's wheelbase for a smooth ride.

It was a perfect showcase for the new science of streamlining, which was being pioneered by aircraft and Zeppelin designer Paul Jaray, whose concepts were licensed by Tatra (and other automakers). A relatively short front end was combined with an arched roofline that gracefully sloped into a long fastback tail. When integrated fenders and a full undertray (belly pan) were added, wind resistance was dramatically reduced. As a result, a relatively modest 60-bhp engine was able to achieve excellent performance and low fuel consumption. A prominent rear dorsal fin ensured high-speed stability. While not pretty like the Art Deco creations of the great French coachbuilders, Tatra's cars were striking and considered, for the period, very modern.

Tatra's radical design philosophy culminated in the production Model T77, which melded Hans Ledwinka's advanced engineering with a smooth and pure teardrop form as displayed in Jaray's earlier work. Presented to the press in Berlin in March 1934, the radical 2.9-liter, V-8-powered machine was arguably the first production car to take advantage of effective streamlining. The T77 was expensive, but less costly versions were planned. It was followed by the Model T77a, which had a 3.4-liter V-8 producing 70 bhp, improved cooling, a three-headlamp configuration, and a top speed of 150 kph. Some T77a's had the center headlamp linked to the steering mechanism. One hundred sixty-seven examples of the T77a were sold. Production of the model ceased in 1939.

Hans Ledwinka understood that the 'tail-heavy' T77/T77a had a tendency for its rear end to swing out in higher-speed corners, so his new, more affordable model, the T87, which first appeared in 1936, had a smaller, magnesium-alloy, 2.97-liter air-cooled SOHC V-8 that, with other refinements,

resulted in a 24 percent (430kg) weight savings. Roadholding was improved, but despite locating the two spare wheels and the battery in front, the T87's weight distribution was still 37 percent front, 63 percent rear. A top speed of 160 kph (100 mph) was attainable, and the model's tendency to oversteer was still a negative, but few automobiles in that era were capable of such effortless high-cruising speeds on the newly developed German autobahns.[2]

In 1937, Tatra introduced the smaller, even lower-priced T97, with a four-cylinder, air-cooled, rear-mounted engine. Just 510 units were made over three years. When the Germans annexed Czechoslovakia in 1938, production of the T97 was prohibited because of its similarity to the KdF-Wagen, which would evolve into the VW Beetle. When the war ended, Tatra sued VW for copying its basic design and won. Compensation was subsequently paid to the Czechoslovakian carmaker. The T87 continued to be produced until 1950. It was followed by the smaller Tatra Model 600 Tatraplan 1946 to 1952.

Tatra T77s, 77a's and T87s are listed as "Full Classics" by the Classic Car Club of America. A 1948 T87 is on permanent display at the Minneapolis Institute of Art. A well-received Tatra class at the 2014 Pebble Beach Concours d'Elegance reacquainted enthusiasts with the innovative brilliance of this pioneering European marque.

—KNG

1   Nick Georgano, ed., *The Beaulieu Encyclopedia of the Automobile*, Vol. 2, (London: The Stationery Office, 2000), 1573.

2   Ivan Margolius and John G. Henry, *Tatra, The Legacy of Hans Ledwinka*, (Veloce Books: Dorchester, UK, 2015), 128-129.

Tatra photos by Peter Harholdt

1939 PANHARD & LEVASSOR TYPE X81 DYNAMIC SEDAN

COLLECTION OF PETER AND MERLE MULLIN

The French firm of Panhard & Levassor was one of the earliest successful automobile manufacturers, having begun in the 19th century. Half a century later, even as war clouds threatened Europe, Panhard & Levassor looked into the future to produce a dramatically different sedan that captured a great deal of attention, thanks to its streamlined styling and innovative passenger configuration.

With the release of its line of *Panoramique*s, Panhard & Levassor set out on a singular path. While most coachbuilders of the time ended up producing pseudo-modern or faux aerodynamics, with a few tilted external surfaces or rounded corners, P&L moved to the head of the pack with the well-received introduction of its new *Dynamic* in Types X76 and X77 at the 1936 Paris Auto Salon. Panhard & Levassor stretched even further in the true spirit of aerodynamic styling with its new X81 two years later, at the 1938 Paris Auto Salon. With a radically new and modern look that emphasized the company's strengths and individualistic approach, the model lineup for the *Dynamic* series separated itself from all other offerings on the market that year.

The X81 had all-new factory coachwork that differed greatly from its predecessors. A fresh, modern design melded the fenders into the body in a sweeping fashion. A new radiator grille leaned fashionably rearward into the car, elegant covered headlamps were faired into the coachwork, the side windows had very rounded upper corners that followed the shape of the roof, and the entire car benefitted from a complementary assortment of soft, flowing curves.

P&L's updated *Panoramique*-style coachwork included an ingenious feature that considerably improved visibility: the windshield A-pillars (support posts) were very slender, framing small glass panes on each corner that helped eliminate the blind spot created by a traditional A-pillar. The side windows also presented outlines similar to those of the *Panoramiques*.

With the new *Dynamic*, Panhard & Levassor displayed its very personal response to the public's frenzy for everything new, modern, and aerodynamic. As first introduced, this large sedan stood out thanks to a peculiar novelty: its centrally located driver's position. The steering wheel was anchored in the middle of the dashboard, the driver flanked by a passenger on either side. This positioning was intended to ensure optimal visibility, a recurring concern at Panhard & Levassor.

The *Dynamic* coachwork, conceived under the supervision of in-house stylist Louis Bionier, displayed shapes that differed radically from those of all other contemporary automobiles. The design of the fenders was uniquely ahead of its time. Eminently French-curved shapes were repeated front and rear, partially covering the wheels. The grille was suggestive of a shield, a motif that was repeated where the two headlamps were faired into the coachwork. The *Dynamic* further asserted its modernity with coachwork that was electrically welded to the chassis to create a rigid unibody.

Four body types were included in the line: a classic sedan, a two-window coupé, a four-window coupé or coach, and a convertible. These various designs were mounted on three wheelbase versions: a 2.60-meter chassis for the *Junior* coupé, a 2.8-meter chassis for the *Major* Coupé and cabriolet, and a 3-meter chassis for the classic and raised sedans.

The 1939 *Dynamic* line featured the the 2.8-liter Type X81 and the 3.84-liter Type X82, both of which had 6-cylinder, sleeve-valve engines. The most important change from previous models pertained to the steering column, which had, by popular demand (and sales that were less than anticipated) been moved to the left side of the car. This alteration helped promote *Dynamic* sales in other countries, as left-hand drive was becoming the new world standard.

Panhard & Levassor ceased production of all *Dynamic* models with the onset of World War II. After the war, as the line had not sold well and the company was struggling financially, it was never resurrected. This example is built on the 3-meter-wheelbase chassis and cost 59,850 French francs upon initial purchase. The top speed with the 6-cylinder engine was 125 kpm, which was respectable for a grand touring car of the period. This handsome *Dynamic* sedan, chassis 222395, is on loan from the Mullin Automotive Museum.

—**RSA**

Panhard & Levassor photos by Michael Furman

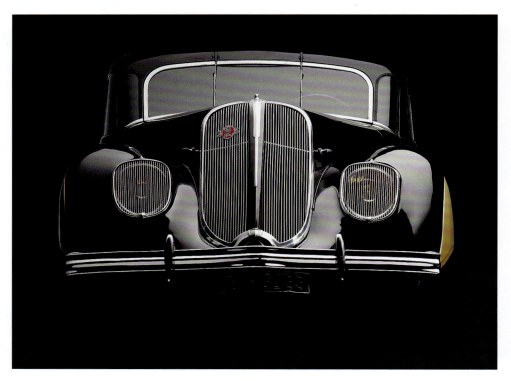

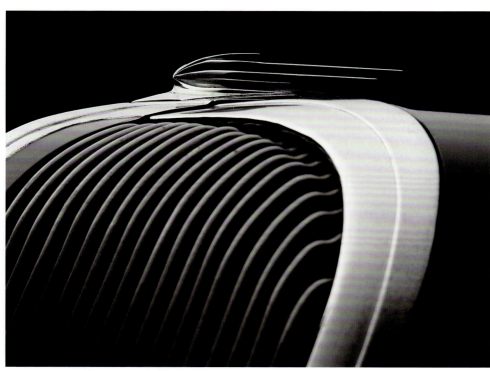

1939 STEYR 55 "BABY"

COURTESY SCOTT R. BOSÉS AND CELESTA PAPPAS-BOSÉS

Streamlined European automobiles in the 1930s were not limited to the expensive makes. By mid-decade, in France, Germany, Austria, and Czechoslovakia, nearly every automaker was experimenting with aerodynamic shapes. The public wasn't yet clamoring for a bolder styling approach, but that hardly stopped Ferdinand Porsche, Hans Ledwinka, Karl Jenschke, and others from designing cars with curvaceous, even flowing lines across a wide price range.

This broad experimentation had no specific formula. Some engineers espoused rear-mounted, even horizontally opposed power plants. They could be air-cooled or water-cooled.

built similar-looking prototypes in Czechoslovakia. These efforts led to the small Tatra V570, perhaps prompting Dr. Porsche to say about Ledwinka, in later years, "Sometimes I looked over his shoulder; and sometimes he looked over mine."

Steyr-Daimler-Puch AG was definitely competitive with VW and the other makes. Its 1936 Model 50, popularly known as the Steyr "Baby," and its later 1938 Model 55 were well-received. The Model 50 Baby's rotund, streamlined shape had been approved for production by Steyr Chief Designer Karl Jenschke in 1935, even though in November of that year he'd moved on to Adler (itself a large proponent of streamlined design) in Frankfurt am Main.

included a folding sheetmetal roof and hydraulic brakes (Volkswagens still had cable-operated brakes). The wheels were stamped steel with holes drilled for cooling.

While the Baby's sleek and smallish, rotund body was reminiscent of several competitors, its powerplant was not. The Model 50's liquid-cooled, horizontally opposed four-cylinder engine relied on the thermosiphon process, so there was no water pump. It was front-mounted, low inside the small engine bay, and it drove the rear wheels through a conventional four-speed manual transmission. To save weight and space, a Dynastart starter/generator powered a small cooling fan. The cars weighed between 1,650 and 1,800 lbs.

The updated Model 55 bowed in 1938, but production ceased in 1940 after the war began. About 13,000 Steyr Baby models were sold between 1938 and 1940, but the Steyr facilities in Austria were bombed heavily.

Unlike the Volkswagen Beetle, which hit its stride after the war, Steyr did not resume Baby production when hostilities ceased. Instead the company built more conventional cars, powered by inline 4s and 6s, as well as the Puch 500, based on the Fiat 500. Rarely seen today, the Steyr Baby is a reminder that streamlined cars spanned the gamut of automobile offerings, from opulent to pedestrian, and from large to very small.

—**KNG**

The [Steyr Baby] compared favorably with the Volkswagen, and outpointed it in several categories.

Chancellor Adolf Hitler's insistence that every German citizen should have an affordable car, and automakers' desire to be the producer of that car, led manufacturers to consider all manner of configurations.

Not surprisingly, some of the most ingenious recommendations for a "People's Car" were small vehicles. The Volkswagen "Beetle" design, powered by a rear-mounted flat four, was one of several similar ideas. Joseph Ledwinka

The public first saw the fastback Steyr Baby at the 1936 Berlin Motor Show. Its rounded front fascia resembled a miniature Chrysler Airflow. The headlights appeared to be high because the car itself was so low. The Model 50 compared favorably with the VW and outpointed it in several categories. Despite its shorter length, the Model 50 offered better seating, suicide front doors, and more luggage capacity than the Beetle. Other differing features

Steyr photos by Peter Harholdt

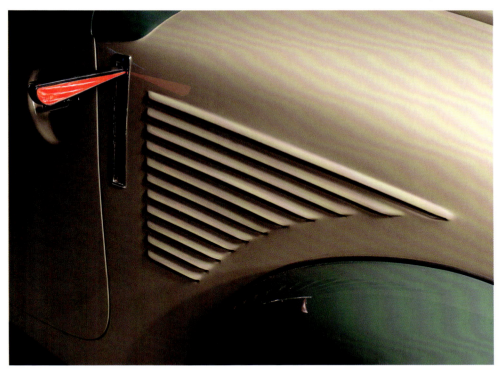

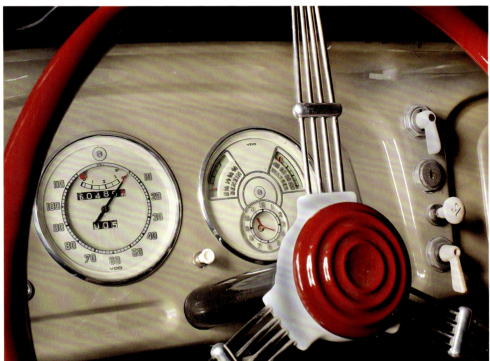

1941 CHRYSLER THUNDERBOLT

COURTESY OF THE RPW COLLECTION

Visitors to the New York International Auto Show in October 1940 were thunderstruck when they gazed upon two sleek convertibles displayed on the Chrysler stand. Five years earlier, Chrysler's then-radical Airflows had failed to move a sufficient number of buyers into a brave new realm of streamlined design. Undeterred, the company briefly dialed back its peek into the future, only to present an even more dramatic approach in Gotham, dazzling the public with a hint of what could be.

With the Thunderbolt Roadster and its four-door companion, the Newport Phaeton, Chrysler offered a distinctly futuristic alternative to conventional models. The company let it be known, discreetly, that while these show cars were ostensibly display pieces, customers of means could purchase one after they had been shown around the country.

Chrysler confidently touted the Thunderbolt as "The Car of the Future." Sporting a smooth, aerodynamic body shell, hidden headlights, enclosed wheels, and a retractable, one-piece metal hardtop (an American first), the roadster was devoid of superfluous ornamentation, with the exception of a single, jagged lightning bolt on each door. It stood apart from everything else on the road, hinting that tomorrow's Chryslers would leave their angular, upright, and more prosaic rivals in the dust.

As if the audacious Thunderbolt two-seater wasn't enough, the four-door Newport phaeton teased all who saw it. With its elegantly stretched proportions, twin leather-lined passenger compartments, artfully dipped doors, and fold-down windshields, the Newport was reminiscent of the classic dual-cowl phaetons popular with well-to-do consumers until the mid-1930s; it was a last attempt to save this retro design concept. Named for the Rhode Island

enclave that was home to extravagant summer "cottages" for the wealthy, the Newport epitomized, in the words of noted auto historian Beverly Rae Kimes, "retro design, 1940s style!"[1]

Adding to the cachet of the Thunderbolt and Newport, the show cars were built by LeBaron, formerly a noted independent custom coachbuilder, now operating as an in-house affiliate of the Briggs Manufacturing Company. In addition to supplying completed production car bodies for makes such as Ford, Packard, and Chrysler, Briggs also offered an independent design consultancy for its clientele. LeBaron's chief designer, Ralph Roberts, who had been responsible for some of the most elegant custom-bodied creations of the classic era, led the creation of the two dream cars. Roberts's co-designer on the project at Briggs was the talented young stylist Alex Tremulis.

The first sketches of the Thunderbolt, developed in 1939, intrigued not only ultraconservative Chrysler Corporation president Kaufman Thuma "K. T." Keller, but also David Wallace, who ran the Chrysler Division. Conventional wisdom held that Chrysler had gone too far with its Airflow notion. Tremulis astutely suggested that these cars be touted as "new milestones in Airflow design," publicly hinting that had it not been for the 1934 Airflows, Chrysler styling might not have evolved into these bold, futuristic shapes. After approving development of both concepts, Chrysler brass insisted they be ready for the 1940 auto show season, in time to debut in New York.

LeBaron employed traditional techniques for limited manufacturing. Aluminum body panels were stretched over wooden framing; the Thunderbolt's full-width hood, which flowed uninterrupted from the base of the windshield to the slender front bumper, and its broad decklid were both steel,

as was the folding, one-piece steel top. Underscoring the Thunderbolt's sleek shape, its curved, one-piece windshield was devoid of vent wings. The side windows were operated hydraulically, and the doors were push button–actuated. Inside, a full bench seat accommodated three passengers. The Art Moderne-design theme extended to the lettering on the instrument dials, with slender white characters set against a light gray background and accented by a bright red needle.

Fluted, anodized aluminum lower-body side trim ran continuously from front to rear. Removable fender skirts covered the wheels, which were inset in front so they could turn. Wind tunnel experiments proved the shrouded wheels were aerodynamically effective, but other than the postwar Nash Airflyte, rival manufacturers (and Chrysler) eschewed this feature. The engine was Chrysler's 140-hp, 323.5-cid straight eight, coupled to an experimental fluid-drive, semiautomatic overdrive transmission.

Priced at a then-lofty $8,250, eight Thunderbolts were planned; only five were built, of which four survive. These dramatically modern cars were well-received, but their high price tag was a deterrent to sales. The onset of World War II meant that, while a few of the Thunderbolt's many advanced features found their way into production DeSotos and Chryslers, these unique show cars were not replicated when hostilities ceased. The present owner, Roger Willbanks, saw this same Thunderbolt as a young boy and was determined someday to own it.

—KNG

NOTES
1 Jonathan A. Stein, *Curves of Steel: Streamlined Automobile Design* (Philadelphia: Coachbuilt Press, 2009), 58–59.

Chrysler Thunderbolt photos by Michael Furman

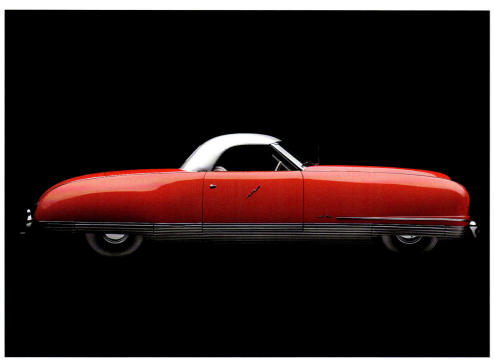

1942 ALFA ROMEO 6C 2500 SS BERTONE BERLINA

COURTESY COLLEZIONE CORRADO LOPRESTO

Alfa Romeo, one of Italy's greatest marques, dates from 1910 in Milan. In 1915, industrialist Nicola Romeo took over the company, then known as A.L.F.A., (Anonima Lombarda Fabbrica Automobili), and the cars were called Alfa Romeo after World War I. Some of the greatest Alfa Romeo racing and road-going models were built in the period before World War II. Alfa won every Mille Miglia open-road endurance race from 1928 to 1938, except for 1931. That era saw the success of Scuderia Ferrari, the company's powerful racing team, managed by Enzo Ferrari, with immortal drivers like Tazio Nuvolari, Achille Varzi, Giuseppe "Nino" Farina, Luigi Chinetti, Sr., and Giuseppe Campari.

Alfa Romeos, at the time, rivaled the finest sports models from Bugatti, and the company's impact could be likened to that of Ferrari today. Alfas in that period included the sublimely sexy, supercharged 8C 2300 and 8C 2900 classics, engineered by the incomparable Vittorio Jano. In 1939, Alfa Romeo introduced its sporty Type 256, powered by a 2.5-liter DOHC I-6 and marketed as the 6C 2500. It was offered in cabriolet, coupe, berlina (five-passenger coupe), and sedan formats. Despite the outbreak of hostilities in Europe, a few 6C 2500s were built as late as 1943 from leftover pre-war components, but only for high-ranking military and civilian clients.

From 1939 to 1942, Scuderia Ferrari modified several 6C 2500 chassis, shortening the wheelbase from 300mm to 270mm. Beginning with chassis number 915501, the tubular steel frame was reinforced with a crossbar, and these cars, equipped with three-carburetor engines, were designated 6C 2500 SS (for Super Sports). A 6C 2500 SS won the Tobruk-to-Tripoli coastal endurance race in 1939.

This stunning example, chassis 915516, according to Alfa Romeo records, is rated as a Super Sport. Author Andrea Cittadini believes the chassis may be an earlier example, "renewed in the 1940s, to cope with the shortage of materials due to the war."[1] Several other Alfa Romeos, including a 6C 2500 cabriolet by Pinin Farina and a roadster built for a military member of the Von Richthofen family, were also assembled during the war years.

Carrozzeria Bertone of Turin built this car's streamlined, custom aluminum body. It was designed by Mario Revelli di Beaumont and initially delivered to Oreste Peverelli, a car dealer from Como, Italy. The berlina was sold in Switzerland in 1946, subsequently found its way to America, and was eventually purchased and completely restored by Corrado Lopresto, whose collection of Italian sports cars and prototypes is one of the finest in the world. A one-off design, the car's twin horizontal lower grilles (which didn't come into popular vogue until the late 1940s) flank a traditional vertical Alfa Romeo vee'd prow. In this case, the Alfa grille incorporates a discrete shape change that harmonizes with the front of the berlina, while it bisects a slender, single-tube bumper. Cittadini called it "a very elegant design that does not hide the speed skills of the car."[2] The coupe boasts fully independent suspension with twin trailing arms and coil springs in front, and a rear swing axle with radius arms and torsion bars.

Other design elements that were unusual firsts on a pre-war car include retractable door and decklid handles, discreet central taillights, and beautifully sweeping fadeaway fenders (later popularized by the postwar Jaguar XK-120 and the 1946-to-1948 Buick). The berlina's low roofline and tapered tail represent streamlining at its very best, and may have influenced the postwar Bentley Continental. Elvio Deganello commented, "Bertone (established in 1912) has been involved in the entire cycle of the automobile's evolution, distinguishing itself with audacious shapes, for its ability to apply innovative technology, and for its courage in putting cars that seemed impossible to build into production."[3]

Donald Osborne said, "While the Chrysler Airflow and its market failure are well known, what is less recognized are the dramatic and much more attractive bodies designed on streamlined and aerodynamic principles created by Italian carrozzerie in the mid-to-late-1930s. Pinin Farina, Bertone, Castagna, Touring, and Zagato all built competition and road cars that explored creative and visually compelling solutions."[4]

This elegant coupe, built in the midst of a war, remains a fortunate survivor and a wonderful glimpse at the way streamlining and beauty could be artfully melded.

—KNG

NOTES

1 *Best in Show: Italian Cars (sic) Masterpieces from the Lopresto Collection.* Andrea Cittadini, Skira Editore S.p.A., Milan, Italy, 2014.

2 ibid.

3 *Italian Coachbuilders: The Masters of Style.* by Elvio Deganello, Giorgio NadaEditore, 2016, pp. 28.

4 *Stile Transatlantico/Transatlantic Style.* Donald Osborne, Coachbuilt Press, Philadelphia, Pennsylvania, 2017, pp. 30.

Alfa Romeo photos by Carrstudio/Collezione Lopresto

EXHIBITION CHECKLIST AND LENDERS

1930 Henderson KJ Streamline
Collection of Frank Westfall, Ner-A-Car Museum, Syracuse, New York

1934 Bendix SWC Prototype Sedan
Collection of Studebaker National Museum, South Bend, Indiana

BMW R7 Concept Motorcycle
Collection of BMW Classic Collection, Stuttgart, Germany

1934 Chrysler Imperial Model CV Airflow Coupe
Collection of David and Lisa Helmer, Macungie, Pennsylvania

1934 Graham Combination Coupe
Charles Mallory, Stamford, Connecticut

1935 Bugatti Type 57 Aérolithe
Collection of Chris Ohrstrom, The Plains, Virginia

1935 Hoffman X-8 Sedan
Myron and Kim Vernis, Akron, Ohio

1936 Cord 812SC Westchester Sedan
Collection of Kevin Cornish, Zionsville, Indiana

1936 Stout Scarab
Collection of Ron Schneider, Franklin, Wisconsin

1937 Airomobile
Collection of National Automobile Museum, The Harrah Collection, Reno, Nevada

1937 Lincoln-Zephyr Coupe
Collection of Alan Johnson, Portland, Oregon

1938 Delahaye 135M Roadster
Collection of the Petersen Automotive Museum, Gift of Margie and Robert E. Petersen Foundation, Los Angeles, California

1938 Mercedes-Benz 540K Stromlinienwagen
Mercedes-Benz Museum, Stuttgart, Germany

1938 Talbot-Lago T-150-C-SS
Collection of Mullin Automotive Museum Foundation, Los Angeles, California

1938 Tatra T77a
Collection of Helena Mitchell and John Long, Toronto, Ontario, Canada

1939 Panhard & Levassor Type X81 Dynamic Sedan
Collection of Peter and Merle Mullin, Los Angeles, California

1939 Steyr 55 "Baby" Coupe
Scott R. Bosés and Celesta Pappas-Bosés, La Canada, California

1941 Chrysler Thunderbolt
Roger Willbanks, Denver, Colorado

1942 Alfa Romeo 6C 2500 SS Bertone Berlina
Collezione Corrado Lopresto, Milan, Italy

ABOUT THE CONTRIBUTORS

Ken Gross (KNG) has been an automotive journalist for more than forty years and has curated several museum exhibitions across the United States, including "The Allure of the Automobile," at the Portland Art Museum. He is the author of *Hot Rod Milestones, The Art of the Hot Rod, Hot Rods and Custom Cars,* and *Speed: The Art of the Performance Automobile,* among others. He is based in Virginia.

Peter Harholdt is a museum photographer and veteran race car driver, based in Florida. Among his recent automotive books are *The Allure of the Automobile, The Art of the Hot Rod, Can-Am Cars in Detail, Museo Ducati,* and the Stance & Speed Monograph Series.

A long-standing love of cars led **Michael Furman** to the challenge of shooting cars in the studio. His work has appeared in several books created in collaboration with Richard Adatto, a fellow contributor to this volume, and the Mullin Automotive Museum, a lender to this exhibition. He is a native of Philadelphia.

Photography has been part of **Dale M. Moreau's** life since the age of five. His father was a commercial photographer who took him on photo assignments and planted the seed that gave him his life's work. The love of the automobile has been the icing on the cake. He is from Salem, Oregon.

Scott Williamson specializes in automotive and vehicle-related photography. He has worked for clients in the OEM automotive industry, custom & hot rod builders, advertising agencies, and marketing firms during more than 25 years of work as a commercial photographer. He is based in Southern California.

Based in Reno, Nevada, **Jeff Dow** has worked with Porsche North America and has shot photos and videos for international banks and local distilleries, famous rock stars and unknown nurses. He has shot on mountain tops and counter tops, from 20,000 feet and below sea level, in 20 countries and six languages.

Montreal-based photographer and writer **Joe Wiecha** is also an award-winning documentary filmmaker, with directing and writing credits on many cable TV networks.

For more than 35 years **David Rand** has been a creative designer working in the automotive industry. He started his career with General Motors, last serving as the executive director of Global Advanced Design. Currently he is a lecturer and design consultant. He is also a judge at numerous concours events around the country and serves on the board of the Concours d'Elegance of America.

Richard Adatto (RSA) is a renowned expert on pre-World War II French cars. he has written or co-authored numerous books, including *Curves of Steel; The Art of Bugatti: Mullin Automotive Museum;* and *French Curves: Delahaye–Delage–Talbot-Lago.*

Jonathan A. Stein (JAS) is known for his work at Hagerty magazine and Automobile Quarterly. He has written three books on automotive history, and his articles have been published in most major motoring magazines.